Mark P. Friedlander, Jr. is an attorney and a writer who has authored and co-authored highly praised works of fiction and award-winning non-fiction for adult and young adult readers on a wide range of subjects from science to history to aviation and business – *When Objects Talk, Outbreak, The Immune System, The Shakespeare Transcripts, Di Vinci Gets a Do-Over, Higher, Faster, and Further* and *The Handbook of Successful Franchising*. His fiction is fun and his non-fiction informative.

This book is dedicated to my wife, Dorothy

Mark P. Friedlander, Jr.

Everybody's Antibodies

Understanding Your Immune System in the World of Covid

Austin Macauley Publishers™
LONDON • CAMBRIDGE • NEW YORK • SHARJAH

Copyright © Mark P. Friedlander, Jr. 2024

All rights reserved. No part of this publication may be reproduced, distributed, or transmitted in any form or by any means, including photocopying, recording, or other electronic or mechanical methods, without the prior written permission of the publisher, except in the case of brief quotations embodied in critical reviews and certain other non-commercial uses permitted by copyright law. For permission requests, write to the publisher.

Any person who commits any unauthorized act in relation to this publication may be liable to criminal prosecution and civil claims for damages.

The medical information in this book is not advice and should not be treated as such. Do not substitute this information for the medical advice of physicians. The information is general and intended to better inform readers of their healthcare. Always consult your doctor for your individual needs.

Ordering Information
Quantity sales: Special discounts are available on quantity purchases by corporations, associations, and others. For details, contact the publisher at the address below.

Publisher's Cataloguing-in-Publication data
Friedlander, Jr., Mark P.
Everybody's Antibodies

ISBN 9781685626204 (Paperback)
ISBN 9781685626211 (ePub e-book)

Library of Congress Control Number: 2023916122

www.austinmacauley.com/us

First Published 2024
Austin Macauley Publishers LLC
40 Wall Street, 33rd Floor, Suite 3302
New York, NY 10005
USA

mail-usa@austinmacauley.com
+1 (646) 5125767

Everybody's Antibodies is a major revision of a previous publication, *The Immune System*, which was published in 1998 by Lerner Publications. The 1998 publication was awarded a place on the 1998 VOYA Non-fiction Honor List and was named Best Children's Book of the Year by Bank Street College in 1999. No second edition was published; therefore, in 2003, all rights to the book reverted to Terry Phillips and me, and then from Terry to me. This current book is a revision that includes all the remarkable and amazing advances in knowledge of the immune system that have occurred during the past quarter of a century.

A special acknowledgment to Dr. Terry M. Phillips, Ph.D., D.Sc., my teacher, consultant, and best of all, my friend.

Table of Contents

Chapter One: In the Beginning: The Starting Gate — 11

Chapter Two: The Protective Wall: Your Skin in the Game — 16

Chapter Three: The Immune System Is Unique: The ABCs of Immunity — 20

Chapter Four: Cellular Immunity: Dancing with the Lymphocytes — 30

Chapter Five: Humoral Immunity: Everybody's Antibodies — 35

Chapter Six: The Functioning Immune System: Lights, Camera, Action — 41

Chapter Seven: Introducing the Invaders: Know Your Enemies — 47

Chapter Eight: Vaccines: Manipulating the Immune System — 65

Chapter Nine: Cytokines: The Multitasking Messengers — 82

Chapter Ten: When It Fails and When It Attacks: The Nightmare — 93

Chapter Eleven: Allergic Reactions: You Get the Itch — 108

**Chapter Twelve: Immune System Nutrition:
 Feeding the Immune Army**   121

**Chapter Thirteen: The Mind-Immunity
 Connection: You Gotta Laugh**   140

**Chapter Fourteen: Immunotherapy: A Look
 at the Future**   154

A Final Note   163

Chapter One
In the Beginning:
The Starting Gate

To begin to understand the marvels of your immune system, you must think small, as in, microscopically small, even molecularly small. Think cells. Think microbes. Think proteins. Think molecules. Today, what we know about the existence of cells, bacteria, and viruses that are invisible to us, except through a microscope, is woven into our common experience. This has not always been the case. Over two thousand years ago, Marcus Varro, the Roman philosopher, wrote about his belief that illnesses were the result of tiny, invisible creatures that were carried through the air from damp and swampy areas and from filth found in populated areas. The memory of his prophetic idea was lost in the mists of time. Instead, for more than twenty centuries, the notion of creatures imperceptible to the naked eye was never seriously considered and was deemed contrary to the accepted wisdom of the times.

The civilized world believed that the cause of all disease resulted from imbalances of the four humors, or body fluids: Earth (black bile from the spleen), water (phlegm from the brain and lungs), fire (yellow bile from the gall bladder), and air (blood from the heart). This concept, known as "humourism," was maintained, and taught, by learned men such as Aristotle, Hippocrates, and Galen. Part of the medical practice of this ancient concept, of keeping the four humors in balance, was bloodletting, a procedure performed by cutting into a vein in the patient's arm and allowing the blood to pour out. It was believed that bloodletting helped keep the humors in balance. Bloodletting was also performed by placing leeches (a bloodsucking, wormlike creature) on the patient's skin to draw out his blood. The fact that people actually survived bloodletting is remarkable, although the number of those who died from loss of blood compared to the number of those who were cured from their illnesses, as a result of this treatment will never be known. Today it is clear that was not good medicine.

The realization of the existence of unseen living creatures began when Antonie van Leeuwenhoek, a Dutch cloth merchant whose hobby was grinding lenses to magnify the image of the material he inspected, decided to use his newly improved lenses in the form of a crude microscope to inspect water. He wanted to see if he could find any differences between rainwater, well water, and water from melted snow. To his surprise, he discovered tiny, wriggling creatures in each of the water samples. His finding that there were "things" squirming around in the water that people drank was revolutionary and a concept difficult to grasp. Van Leeuwenhoek called these creatures "wee animalcules." Using his new lenses, he continued to study these small creatures, as well as larger ones like weevils and fleas. His studies soon gained the attention of the Royal Society of England. Although the "wee animalcules" he had observed were actually bacteria, neither he, nor any member of the Royal Society with whom he exchanged correspondence for almost fifty years, ever connected these "wee animalcules" with disease.

However, the significance of his discovery—what the animalcules did and whether they were good or bad—was not fully understood until 1857, when the French chemist, Louis Pasteur, established that the wee animalcules (bacteria) could, in fact, cause disease. It was this revelation that, according to some authorities, resulted in Pasteur's designation as the father of immunology and, certainly, the father of bacteriology. On the other hand, there are many who attribute that honor to Edward Jenner, an English doctor who, in 1796, developed a smallpox vaccine by pre-infecting patients with cowpox, a mild disease, to block infection from the highly contagious (and deadly) smallpox. Jenner's vaccine caused a person to become immune to the disease.

Our journey into learning the basics of immunology begins with cells. Cells are the basic building blocks that make up all living creatures. You are made up of billions of cells that are grouped together into tissues and organs. When you think of billions, think of how enormous this number is and then understand how tiny your cells must be to fill your body with billions of cells. Simply put, if your cells were the size of marbles, one billion of them would fill your school gymnasium; or, if you wanted to count each cell (one second for each number), it would take you thirty-two years, counting all day and night, nonstop. With that in mind, let's put size into perspective. Cells are so small that ten thousand of them could fit on the head of a pin. And then picture molecules, such as proteins: two million (2,000,000,000) would fit on the head of that very same pin. It's hard to comprehend, isn't it?

Your cells are the product of the joining together of genetic material contained in cells from your mother and cells from your father. Each cell consists of molecular units (i.e., genes) that determine the characteristics you inherit from your parents. Genes are the cell-control mechanisms contained inside the nucleus, the largest component of each cell.

When you were born, your immune system was not quite ready to take on the sea of germs—sometimes called microbes, microorganisms, or pathogens—in which we all live. In the normal course of a day, you are exposed to germs in the air you breathe, on the things you touch, from the people you are with, and in the food you eat or the beverages you drink. A baby is protected, as much as possible, by its parents and the sanitary conditions they provide. Mother's milk also provides protection since it contains immune proteins (antibodies) from her immune system that helps the baby fight any invaders.

Deep within the soft marrow of the long bones of your legs, arms, and your breastbone plate (the "sternum"), are important development cells called stem cells, which produce millions and millions of other types of cells every day, including red blood cells, platelets, and white blood cells. Red blood cells carry food and oxygen to all the other cells of your body and remove carbon dioxide and other cellular waste. Platelets provide clotting to help heal cuts and wounds. White blood cells are your immune system guardians – the protectors against body invaders.

By "body" invaders, we mean anything that is not a natural part of your body, such as bacteria, viruses, parasites, fungi, pollens, and even transplanted organs. How the white blood cells work, what they do, and the molecular network associated with them are the subjects of this book. Understand that although each part is discussed separately, they all work together in a complex, intricate, and mingled network of activity.

Chapter Two
The Protective Wall:
Your Skin in the Game

You need to have "some skin in the game." The expression "skin in the game" probably came from the long-forgotten card game "Skin" or as some believe, a reference from owners of derby racehorses, or from the financier, Warren Buffett's reference, of putting his own money into an investment as "skin in the game." The relevance of this expression simply gives emphasis to the importance of our skin. What does that expression have to do with understanding how your immune system works? It is used here to call your attention to the fact that your skin is the first line of defense in protecting your body from body invaders.

The world we live in is alive with 250 million different types of germs. They are everywhere: On the things we touch, on what we eat, and in the air we breathe. These germs—pathogens, microbes—by whatever name we call them, are all potential invaders: Viruses, bacteria, protozoa, and fungi, anxious to find ways to invade our bodies, your body, and set up their own disease-causing neighborhoods.

In the invisible sea of the thousands of germs that surround you every day, your skin (the body's remarkable coating) is your dike: Your walled fortress of protection against viruses, bacteria, protozoa, and fungi.

Your skin, the whole-body wrap, is a single organ that is tough and stretchy. With its three layers, your innards are guarded from microscopic attacks: The deepest layer is the subcutaneous fat hypodermis, which cushions the skin layers and lies between the skin and the muscles. Next comes the dermis, a layer containing blood vessels, lymphatics, and nerves that regulate your body's temperature. The dermis has a "basement" layer that produces the cells that make up the skin. The dermis is the main layer where most of your skin's functions are performed, such as sweating, feeling sensations, hair follicles growing, and oils are created to prevent dryness. This is also the layer where issues may present: Acne, pimples, and zits. The top layer is the epidermis, the thin, final layer, or waterproof jacket, that is the exterior wall, defending against pathogenic invaders. This layer is made up of dead and dying cells from the dermis and acts as a frontline barrier.

In general, no microbe, or chemical, can penetrate healthy skin, although, unfortunately, there are some exceptions, such as poisons like formaldehyde and poison ivy oils. The natural body openings through the skin (such as the eyes, ears, nose, mouth, vagina, and anus) provide possible ports of entry. Fortunately, these orifices are protected by mucosa, tissue that provides immunity and specialized fluids such as tears, saliva, and mucous, a thick fluid that binds up the invaders so they can be eliminated. In the moisture of these surfaces are specialized cells and antibodies (we'll meet them later) that recognize and attack disease-producing microbes. Cuts that penetrate the skin or burns that remove layers of the skin may offer the body invaders an entrance for attack.

The thousands upon thousands of germs that are usually on your skin are regularly removed when you bathe. Washing your hands is an obvious protection: The structure of soap is similar to the fat structure of the virus coat, enabling it to interact and damage, or disrupt, the surfaces of viruses and bacteria. When the suds are rinsed away under running water, the damaged virus and/or bacteria are also removed. This is why the CDC recommends washing hands for at least thirty seconds to remove germs. This has been particularly important during the COVID-19 pandemic. You also want to be conscious of the fact that whenever you touch your eyes, nose, or mouth, you are offering the waiting microbes an inviting passage past the defense of your skin wall.

As we proceed together through these chapters, we will look at each section separately, but as you will hopefully understand, all the parts work in harmony in the body's intricate, mingled, complex web of fighting soldiers in your service.

Chapter Three
The Immune System Is Unique: The ABCs of Immunity

Your immune system is comprised of millions and millions of white blood cells working with billions of smaller Y-shaped protein-molecules (which we will discuss in more detail in Chapter Five) called antibodies and other molecules. These cells communicate by way of a complex web of signals. The organization of these components of your immune system can be visualized as a cellular and protein molecular army with infantry, artillery, and reconnaissance units, all operating from a network of command centers called "lymph nodes." All of these cells and protein molecules travel through the body and around the organs of your body, riding and plunging through the rivers formed by your blood vessels to merge into the streams and creeks formed by your lymph system and intrude into the capillaries, as rivulets flowing between all the cells of your organs, such as your heart, lungs, liver, brain, and skin cells.

Your immune system responses come in two basic types of white blood cells: The first type will attack any invaders they find as they flow constantly through the rivers and streams of your blood and your lymph system, and the second type is comprised of those that attack specific invaders. The white blood cells that attack any hostile intruder are like front-line soldiers always trying to defeat an enemy themselves if they can. The second type of white blood cells that attack specific invaders are "lymphocytes," which can develop into T cells, and B cells, capable of producing the Y-shaped antibodies; both types of cells can be directed to handle a more dangerous intruder.

While all white blood cells are generated in the bone marrow of your long bones or breast plate, they become modified as your body's needs are determined. The white blood cells that pass through a gland under the breast plate—the thymus, a glandular tissue—are given further modification where they are refined into different types of T cells, such as helper T cells, suppressor T cells, killer (or Effector) T cells, or natural killer (NK) T cells. B cells remain in the bone marrow and mature there. We shall meet each of these white blood cells as we look further into the immune system units.

As an example of the army-like organization of your immune system, when a front-line cell called a "macrophage," the largest of the white blood cells, locates a body invader, it immediately attacks. When it is joined by more white blood cells, a message is sent through molecular messengers—cytokines, interferons, and growth factors—to the lymph nodes, the headquarters for the immune command. There, the white blood cell generals, the helper T cells and the suppressor T cells, direct other immune units to the fight. As we shall discover in later chapters, the battle plans are complex and involve various levels of immune warfare. T cells form the infantry, cells that specifically seek out and attack their targets, while the B cells are the artillery units that are the source of the molecular rockets called "antibodies," the force that attacks specifically identified germs.

Key Factors

There are three key factors that make the immune system remarkable:

1. <u>It distinguishes</u>. Your immune system can tell one attacking invader from another. For instance, if a mumps virus enters your body, your white blood cells (macrophages and accessory cells) recognize it as a mumps virus and not as bacteria, a fungus, or even, for instance, a different type of virus, such as a flu virus. After the invader is identified, a report is sent to the command center to tell the white blood cell generals at headquarters what type of invader is present so that the appropriate response teams (i.e., killer white blood cells,

antibodies, or both) can be sent to the site.

2. <u>It remembers</u>. After your body has been infected once with a disease-causing germ, such as the previously mentioned mumps virus, your immune system will remember that the mumps virus is a bad guy. At the same time, it will remember the correct action to take, like sending out either anti-mumps killer T cells or commanding the B cells to make anti-virus antibodies. The anti-virus antibodies will then be called to duty to rocket in and destroy the mumps virus invader, before that virus can get a head start on causing an illness. The fact that your immune system has a memory makes vaccines work effectively to protect you without you being aware that a pathogenic invader has attacked and has thus been destroyed, meaning that vaccines work without our conscious knowledge.

3. <u>It knows itself</u>. When you were born, every cell of your body was tagged with an identical and special molecular set of identification markers or badges. These badges are called "human histocompatibility antigens" (HLA). Everybody's badges are unique to them meaning they are different from any other person's badge. Just like human fingerprints, some are vastly dissimilar from other people's

badges, and some are only slightly distinctive. The infantry patrols (white blood cells) flow through your body looking for invaders; they know the neighborhood, and because they recognize all your body cells, they do not attack your own body. This is why patients who have organ transplants sometimes experience problems. In organ transplants, the transplanted organ is not recognized as containing cells that belong to your body, so in order to survive, the immune system must be medically suppressed.

For example, when the immune system's white blood cells see a new transplanted organ, they say, "Hey, that wasn't here before. That organ has never been part of this body – its molecular badge does not check out. We have to attack it." It is this part of the immune system that must be understood and addressed with the use of immunosuppressive medications before organ transplantation can be performed, which is why doctors seeking organ donors for transplantation have to run special tests to check for the human leukocyte antigen (HLA) in potential donors. This is called tissue typing. If there is a "close match," the potential organ donation can occur after the patient has been administered immunosuppressive medications. When the "match" is not close, the organ transplant cannot take place because partial, or total, rejection by the immune system would result. The immune-system soldier would view the identification badges and say, "Sorry, you are not allowed to stay here."

The vast and complex army of the immune system is divided into two distinct commands. The cells that are the front-line defenders are white blood cells that can engulf and destroy any invaders, and they, together with the physical barriers (skin, tears, saliva, and sweat), form the "innate" immune system. The second command is the specific T cells and B cells (plus antibodies) that make up the more powerful "acquired" immune system. The acquired immune system is further divided into the white blood cells that travel through your body, acting as a motorized infantry, or soldiers with boots on the ground, and the generals are at the command centers. There are many different types of T cells, each with a separate task. Some attack foreign invaders wherever they are, while certain T cells attack specific invaders and are backed up by the white blood cells that form the B-cells, which produce specific antibodies against identified foreign invaders.

Types of Immunity

Innate immunity is the first line of defense within your body against any foreign invader that gets past the barbed wire wall (your skin). In some cases, this is the only type of defense needed. The second, more complicated, line of defense—the acquired immune system—is divided into two different branches: "cellular" and "humoral" immunity. Protection requires a combination of these two branches; a combination of specific, or primed T cells (cellular), and B-cells, which produce antigen-specific antibodies, which together, with a series of special proteins (complement), help the antibodies destroy identified invaders. Billions of antibodies are generated to fight specific invaders. Antibodies and the white blood cells (B cells) that manufacture them make up what is called "humoral" immunity. These names come from how the two different systems work – cellular being named because it is comprised of cell components and humoral immunity is named because it involves substances found in the humors, or body fluids.

Some immunologists maintain that cellular immunity is the major immune fighting force, while others believe that humoral immunity, a powerful and intricate system producing antibodies, is the major fighting force. However, antibodies are often ineffective against intracellular invaders which can only be destroyed by specific T cells, so they all must act together to make the system work. For example, tuberculosis requires cellular immunity to destroy the intracellular infection, while viruses and bacteria can easily be destroyed by specific antibodies. In reality, a combination of all aspects of the immune system, innate, cellular, and humoral, are required to ensure a healthy body.

A simple rule to remember the difference between innate and acquired immunity is that defense mechanisms that DO NOT require lymphocytes are "innate," while mechanisms that rely entirely on lymphocytes are "acquired." As we have explained, there are no clear lines between the army units. In the fog of battle, many components mix and match. And as you will find, as we explain more, there are other components that also play major roles in the fray – in your defense.

The battle begins when the invaders are identified and the white blood cells swing into action. The invaders all contain their own individualized identification badges, which would be like a germ wearing a name tag reading, "Hi. My name is Joe and I am a very bad flu virus." These badges are, in reality, unique molecules on the surface of each invader. These molecules are identified by the circulating white blood cells, which, in turn, trigger your body to begin the battle to protect you.

These molecular badges are called "antigens." Specifically, an antigen is any part of an invader that invokes a response from your immune system. While we expect to find these antigenic badges on recognized invaders such as bacteria, viruses, protozoa, and fungi, there are other invaders like pollens, some chemicals, drugs, and even organ transplants—as previously mentioned—that can initiate an antigenic response from your immune system. On the other hand, nylon, Teflon, the metal in a hypodermic needle, or even an implanted heart defibrillator (that is made of metal) would usually not cause an immune (antigenic) response in contrast to the response that an implanted, donated organ would. The difference is that the organ's exterior cells contain cellular markers that are identified as not belonging, whereas the nylon and metals contain no such markers, and are not considered by the immune system to be foreign or antigenic.

All viruses are antigenic to almost everybody. Bacteria and other microbial invaders, such as yeasts and parasites, are antigenic to most people, although not to everyone.

If you cut your finger with a knife, the metal of the blade is not antigenic by itself and would not cause an immune response. However, the germs or pathogens on the knife and on your skin would enter the cut with the blade and create a rush of white blood cells to the injury. That is why a doctor performing surgery carefully disinfects the skin at the operation site with alcohol or with another cleanser and uses only a sterilized scalpel to ensure that no antigenic invader enters the incision, because a sterilized scalpel is not antigenic.

Antigens also create problems for those having blood transfusions. Blood is typed by antigenic characteristics on the red blood cell's surface, known as the ABO system. For instance, Type A blood has its own molecular antigenic identification badge, while Type B has a different badge. Type AB has antigens of both, while Type O has no badge. A person with Type A blood cannot receive Type B or Type AB blood without invoking a violent immune response. A person with Type A blood can receive Type A blood from another person or from Type O negative blood, which is known as the universal donor because it has no A nor B surface antigenic markers to elicit an immune response, while type AB can receive either type A or B Another important blood group antigen is known as the Rhesus antigen (Rh), and this antigen can be associated with all of the known major ABO antigens. The addition of this factor means that every blood group antigen is further divided into Rh positive (+) and Rh negative (-), depending on whether or not the Rh factor is present. The significance of this factor is that it further complicates blood transfusions because giving O Rh+ blood to an O Rh- recipient can cause an immune reaction known as Rhesus Disease.

Chapter Four
Cellular Immunity:
Dancing with the Lymphocytes

While all parts of your immune system work in harmony, cellular and humoral immunity will be explained separately to better explain each network. Your immune system can be viewed as an army of white blood cells, also called "leukocytes." The foot, front-line soldiers are those white blood cells called "macrophages." These are the biggest of the white blood cells. They flow throughout the body, especially around the skin cells, so that they can deal with invaders as soon as they enter the body. Upon encountering an invader, the macrophage stretches itself and changes its shape in order to engulf and dissolve any potential invader as well as cellular debris. The macrophages will attack any invader without a proper identification badge. Just like other white blood cells, macrophages have many complex functions, but principally they are on the front line of defense. Also, upon encountering an invader, macrophages emit chemical messengers called "cytokines" (which will be discussed later in Chapter Nine) to alert other members of the immune forces.

When the invading force is more than the macrophages can handle, other units are called in. The "neutrophils," the most abundant of the white blood cells, are also among the first responders. They are particularly quick to respond to an invasion by bacteria.

They respond quickly to signals from specific cytokines so that they can rush to the site of an infection. The major cells in pus—the whitish-yellowish material in boils and zits—are, for the most part, neutrophils. Other first responders are called "basophils" and "eosinophils," which are specialized white blood cells designed to respond to invaders such as pollen and parasites. These cells, together with the neutrophils, are collectively called "polymorph nuclear cells," or "PMNs," because of their strange, often tri-lobed, nuclei, combined with the presence of intracellular granules filled with "enzymes." As explained earlier, the immune system is an intricate web; it embodies numerous components that work together, but each element has designated functions, although they have the ability to multi-task. For instance, the eosinophils are primarily responsible for attacking large multicellular parasites, such as worms and other parasites. However, they can also fight alongside basophils and become involved in allergies and asthma (as will be discussed later in Chapter Eleven).

The Main Force

The main force, and most numerous, of the white blood cells are the "lymphocytes." After they are manufactured by the stem cells, some lymphocytes are carried to the thymus, a gland under the breastbone. The thymus is large during childhood but becomes smaller as a person gets older. In the thymus, lymphocytes go through bootcamp to become T cells. In training, they each become specialized in order to perform the many different tasks of the immune system. Some of the white blood cells become helper T cells, some suppressor T cells, some (effector) T cells, and some natural killer (NK) cells.

These cells are the basic infantry combat units. Lymphocytes directly attack any invader that carries an identified antigenic marker. The helper T cells and the suppressor T cells are the generals at the central command centers – lymph nodes. Whenever a cytokine messenger arrives with the news that a macrophage or a lymphocyte, or any type of white blood cell, is being overwhelmed, they order killer T cells to the rescue. The purpose of a helper T cell and a suppressor T cell is to keep all the attack plans in balance. The helper T cell generals are programmed to send more killer T cells than needed while the task of the suppressor T cell generals is to keep the outrush in check. Together, the helper and the suppressor T cells keep a perfect balance so that exactly the right number of troops arrive at battle: If too few are sent, the system would lose to the invaders, and if too many are sent, the deadly killer cells might do harm to your body, unintentionally triggering what is called "an autoimmune reaction," which is when your own immune system attacks your body (which shall be described in Chapter Ten).

When the headquarter is not coordinated and the two generals—helper and suppressor T cells—cannot agree, serious health problems occur. For example, when the virus known as "human immunodeficiency virus (HIV)," the cause of "acquired immunodeficiency syndrome" (AIDS), attacks and destroys only the helper T cells, the system becomes unbalanced; as a result, not enough killer T cells are deployed. This eventually leads to a situation in which no killer cells are sent to stop the invaders. This also means that the person suffering with AIDS has a weakened, or even nonfunctioning, immune system and is, therefore, no longer protected against other diseases, such as pneumonia, which is caused by inflammation of the lungs infected by viruses, bacteria, or fungi. The killer T cells and B cells, in the right numbers, that would have been called into battle by the helper T cells, and which would easily have destroyed the invading bacteria, are not functioning. This creates a dysfunctional immune system. Without the helper T cell general, invaders can gain the upper hand and make that person sick, or even cause death. (See also Chapter 10)

Chapter Five
Humoral Immunity:
Everybody's Antibodies

When an invasion appears to overwhelm the body's defenses, the central command of T cells will call into action another type of white blood cell: The B cell, which will produce antibodies (the molecular artillery) that function as a powerful weapon in the army of the immune system. As mentioned earlier, antibody-mediated immune attacks are known as humoral immunity. Antibodies are protein molecules that are shaped like a Y. At the end of each arm of the Y is a claw-like receptacle, which looks a little like a molecular lobster. In its cluster of molecules looks more like a lobster in a fuzzy coat. Both claws are shaped to fit the shape of one specific antigen found on a specific biological invader, such as a virus, bacteria, fungus, parasite, or any other antigen. Each antibody is manufactured in the modified B cell (the "plasma cell"), where it is produced with a molecular receptor to match, and inserts itself inside an invading antigen, like a key fitting into an antibody lock. For this reason, each antibody reacts with a single invader. The antibody's claw-like receptors and the antigen's surface molecular badges are like two pieces of a jigsaw puzzle that fit together perfectly.

Antibody production begins in the B cells. When called upon by cytokine messages, the B cells transform into plasma cells, which in turn become antibody manufacturing plants.

When the molecular identity badges of the invader are revealed to the B cell, it becomes a plasma cell and manufactures an antibody with a duplicate molecular counterpart that can lock onto the antigen.

Once manufactured, the rocket-like antibodies become a powerful part of the immunological weaponry. As the second (important) line of defense, antibodies are a deadly force against invaders. They zero in on specific, identified invaders, attach to them, and destroy them, with the aid of a series of digesting proteins called "complement."

Each B cell has antibodies on its surface – over 10,000 on each cell. When an invader has penetrated into the body and been identified, molecular messages instruct the B cells to produce antibodies specific for that invader. Two new B cells are generated: The plasma cell, which functions as a factory, manufacturing billions of antibodies that swarm to the attack; and memory B cells that hang around in the lymph nodes to remain in memory as part of an immunological directory. In the event that the same invader should show up again, these B cells are ready to attack immediately, initiating plasma cells and rapidly producing specific antibodies.

Normally, billions of antibodies are processing through your system at all times; therefore, your system is likely to have a wide-enough range of antibodies to be able to identify most invaders and can, by the described process, learn the identity of others. This is why the gradual exposure during your lifetime, particularly during childhood, to many bacteria, viruses, and other invaders helps to develop the natural protection of your immune system. A person growing up in an entirely bacteria and virus free environment will not have the natural immunity of children exposed to the normal pathogens of their natural environments.

For instance, children growing up in a country where certain bacteria are found in the water they drink or in which they wash, could be totally immune to diseases caused by that bacteria. Whereas people in a country whose water does not contain that bacteria would become ill if they drank the natural water found in the first country. This is why many people, when they travel, are careful to drink only bottled, or boiled, water and do not add local ice to their drinks. If you are from the United States and travel to Mexico, you would likely never have been exposed to a number of bacteria, the most common of which is *Escherichia coli* (E. coli), usually found in drinking water in Mexico and not in the water where you were raised. As a result, if you drink the water at a local Mexican restaurant during your travel, you might find yourself suffering from diarrhea, also known as "Montezuma's Revenge."

If you have ever suffered from chickenpox, you will thereafter have chickenpox antibodies ready to attack the virus if it shows up again, and you will be immune to any future chickenpox infections. The same thing happens when you receive a chickenpox vaccination. Without having to suffer the disease, your body is fooled into making anti-chickenpox antibodies on a scale too minor to cause any chickenpox symptoms, but just enough so that the chickenpox virus antigen is now in your immunological memory in the B cells of your lymph nodes and is ready to immediately attack an incoming chickenpox virus whenever it appears. Because of the vaccine, you will never have symptoms of the disease and will be unaware that you were attacked.

Antibodies cannot think, although the B cells that produce them do develop antigen memories. They are simply molecules, but they are magnificent molecules that can attach to their specific, memorized antigens. If an antigen appears on any cell—foreign or self—the antibody will attach to it and, with the help of complement, will drill holes into the cell membrane and destroy the invader, in very much the same way the killer T cells work, or macrophages wrap around an antigenic invader to dissolve it. Your immune system will consider your infected cells to be damaged and thus will annihilate them. The macrophages and neutrophils then act as part of the immune system's janitorial service and clean the debris, which prevents further reactions from occurring. Because producing antibodies takes a lot of energy, your body only calls upon antibodies to fight the major body invaders.

For example, an antibody is called to duty when an overwhelmed macrophage sends a messenger back to the command center with a molecular fragment taken from the invader; this is done to create a template the plasma cells use to copy for manufacturing the necessary antibody. It not only takes energy to manufacture antibodies; it also takes time. The B cells must evolve into plasma cells before they can get the cell factory up and running. It usually takes days, sometimes weeks to get into full production. This is why it sometimes takes time to rid your body of a disease, particularly a viral disease, which normally triggers an antibody response.

Think about a lingering head cold. When a virus finds its way into your nose and respiratory system, you suffer from a cold: A head cold, chest cold, or sore throat. The antibody production begins immediately, but it takes days for enough antibodies to be produced to overwhelm the viruses, which are busily multiplying themselves in their attack on your body.

Antibodies remain active in your body for only three weeks. They are then flushed away to make room for new antibodies. The B cells with stored memory are important because they will remain in the lymph nodes and retain a memory of the antigen they were made to specifically fight. Thus, the immune system remembers past invaders!

Chapter Six
The Functioning Immune System: Lights, Camera, Action

Teamwork is defined by Google as a "combined group of people, especially when effective and efficient," while a dictionary defines teamwork as a "coordination of effort to collective efficiency." These definitions are also a description of a well-functioning immune system. All the units described in the preceding chapters, combined with the messengers and occasional commanders in Chapter Nine, when functioning as a team, are the key to good health, your good health. All the units must act together. For example, when you cut yourself, a tissue reaction is followed by a reaction from your immune system. The type of reaction depends on the kind of cut. If a doctor in a hospital operating room makes a surgical cut in an area of your skin that has been sterilized (cleaned to ensure that no living organisms are present), the skin will become red and tender as white blood cells and platelets rush to the site of the wound. This natural reaction to injury is called "inflammation." But, although white blood cells and platelets come to the site, they will find almost no invading bacteria upon arrival. After the platelets seal the stitched incision with a scab, the white blood cells remove dead cells as a part of the healing process. This can happen within ten to twenty minutes of bringing the cut surfaces together.

If you are working in the garden and prick your finger on a rose thorn, bacteria are immediately injected under your skin, causing its surface to swell and turn red. The first line of defense, the macrophages and neutrophils, surround the invading bacteria and destroy them, before carting away the debris. In a day or two, the redness will be gone.

If, however, you slash your hand and cut all the way through to draw blood, a large number of bacteria will enter through the skin. Your immune system will sound a general alarm and call many different types of white blood cells to action. Not only will the macrophages and neutrophils take part in the attack, but the macrophages will also report the invasion to the T-cell command center. In addition, they will release cytokines to alert the command center. The command center will then send killer T cells to the site and alert the B cells to begin antibody production. The cut will become dark red and painful. This is because the small nerve endings within the skin, and deeper tissue have been damaged, releasing pain-inducing chemicals called "neuropeptides." These chemicals interact with the surrounding tissue to cause inflammation and local tissue swelling. Inflammation further exerts pressure on undamaged nerve endings, which, in turn, become irritated, causing the sensation of pain. The neuropeptides also attract other white blood cells to the area, thus increasing pressure on the nerves. When the swelling becomes reduced, the cut heals, and so the tissue pressure will lessen and the nerves will become normal, diminishing the pain.

Now, should the garden where you cut your hand contain manure-enriched dirt, that would a pose further danger. A bacterium entering through the cut in your hand could be *Clostridium tetani*, the bacteria that causes the disease tetanus. Tetanus bacteria live in many types of dirt, in rust, and especially in manure. You could have a really serious problem, or not so serious, depending completely upon whether, or not, you have had a tetanus vaccination.

If you did not have a tetanus vaccination, those bacteria would quickly overwhelm your white blood cells. Within a few weeks, you would become depressed and have a headache. Then your jaw and throat muscles would tighten, and it would be hard for you to breathe. You could die. But, if you have had a tetanus shot, the antibodies in your B cell memory, aided by the vaccine, would immediately overwhelm the tetanus bacteria and destroy them. Booster shots are given regularly to keep the tetanus vaccine up-to-date. Even if you haven't had a booster shot, receiving a tetanus booster within a day of cutting yourself would give you the ability to destroy all the tetanus invaders. That's because the tetanus bacteria grow slowly, and the booster creates a fast response. By bedtime, only a few of the invaders would remain, but the cut would still be red and sore, as the white blood cells—your janitorial macrophages and neutrophils—cart off the debris of battle, and your tissues begin to repair themselves.

Quickly washing dirt from a cut helps your immune system by preventing additional bacteria from gaining entry into your body. The flow of plain cold tap water works well if it is running in a constant stream. Public water is purified and contains only a small number of bacteria. Flowing water is a good cleanser because bacteria (which have no hooks) are easily carried away in flowing fluids. A gentle wipe with clean cotton, soaked in alcohol or another disinfectant, also destroys bacteria, thereby reducing the chance of greater infection.

The Danger of the Burn

Burns occur when you are injured by heat or fire. Burns may allow bacteria to enter your body. A burn destroys the outer layer of skin and allows lymph fluid to seep through the burned area. When this fluid pools on your skin, it creates a great place for germs to grow. Lymph fluid is rich in food materials normally used to feed the cells in your body, but it is also very appealing to bacteria. At the same time, because a burn seals the blood vessels and lymph vessels in the area, the rivers and streams that normally carry your army of white blood cells to the battle zone become blocked. This gives the bacteria a head start, making it harder for your immune system to fight off infection.

A minor burn quickly blisters, providing a natural shield to stop further bacterial invasion. That is why you should not puncture a blister (or a burn's surface). The skin covering the blister protects you.

A third-degree burn, which is a really serious burn, penetrates all the layers of your skin right down to muscle. The deep burn fissures trap bacteria, and infection is likely. The shock of a burn also weakens your immune army functions. Often, people who have been seriously burned must lie in a hospital bed with their damaged skin uncovered, which helps the skin scab over naturally. These people are usually kept away from all visitors to keep even the natural bacteria that live on a visitor's clothes or skin from seeking haven in the unprotected open burn wound. If the burn is really deep, saline-soaked gauze is often used to cover the damaged areas, creating an artificial skin to help keep out the germs.

The immune system can also be involved in making us sick. Allergies, like hay fever, are caused when the immune system overreacts to pollen, or dust particles, in the air. This is also true for food allergies and reactions to many ordinary materials. Allergies caused by chemicals and oils, such as the oil produced by poison ivy, are different, and the immune response to the skin reactions that take place is caused by attacking T cells.

Autoimmune diseases are caused by an overactive immune system. In the case of rheumatoid arthritis, the damage to the joint tissue is caused by the immune system's cleaning crews following an initial antibody attack. Usually, antibody fragments, together with complement, remain in the joint fluid after the attack, and parts of the complement attract neutrophils and macrophages to the site. This immune debris, called "immune complexes," can slowly become deposited on the linings of the joints, attracting over-zealous white blood cells that, in turn, lead to the damage we know as arthritis.

The immune system as an enemy is detailed in a later chapter (Chapter Ten).

Chapter Seven
Introducing the Invaders:
Know Your Enemies

In the earlier chapters, we referred to the body invaders simply as germs, or pathogens. Now, we want to actually look at those invaders in great detail. There are six major types of body invaders: viruses, bacteria, parasites, fungi, foreign cells, and chemicals.

Virus

A virus is one of the smallest and simplest of all life forms. Viruses are so tiny that they cannot be seen with an ordinary microscope. Only after the development of the electron microscope (EM) was it possible to see viruses. An EM uses an electron beam instead of a beam of light to create an enlarged image of a small object. The electron beam relays the image to a TV screen to allow the investigators to examine their target. The picture on the screen is many thousands of times the actual size of the object. This should begin to give you an idea of how tiny, really tiny, a virus is. Just how small is that? More than a million viruses could fit on the head of a pin. If you could imagine that a bacterium, which you cannot see except with an ordinary light microscope, was the size of a football, then a virus would be the size of a marble.

While viruses come in many shapes and there are thousands of different viruses, most are either ball or cylinder shaped. There are many academic arguments as to whether a virus is living or non-living. Is this important? It is, as we try to understand how viruses differ from other invaders.

As a living thing, a virus possesses genetic material that multiplies into more of itself and can change, or mutate, into new characteristics. Since the manufacture of the flu vaccine must be done in the spring, in anticipation of the fall-winter flu season, the Center of Disease Control must make educated guesses as to the type of flu likely to appear. For instance, a flu virus, for which a vaccine is developed, can mutate into new identifiers, changing the nature of its protein outer shell so that new generations of that flu virus would not be defeated by the existing flu vaccination. This is an example of why every year a careful person should get a new flu vaccination in order to match the new emerging flu virus.

On the other hand, all viruses lack the essential features of living things: They are not made up of cells; they contain only genetic material and a protein coat, and they cannot reproduce themselves outside of a host cell, be it human, animal, or plant. They are small but can be the cause of many dangerous, and often deadly, diseases. Outside of a host cell, a virus is inert, inactive; that is, it can't do much of anything. Instead, the virus must wait for the chance to find the proper person, animal, or plant, then gain entrance, invade a cell, and then take over the cell's control mechanism. Once that happens, the virus comes to life and causes the infected cell to begin producing many copies of the virus.

Essentially, a virus is like a baseball. It contains an outside shell composed of proteins, while its inside is stuffed with strand-like material, known as nucleic acid, or RNA. While some viruses may contain DNA, most contain RNA. Often the virus will enter the human body, perhaps your body, drifting on the droplets coughed, or sneezed, from someone near you, or by traveling on the fingers of your hand that touched an object – a dirty handkerchief, or even a doorknob-- and then floats on your finger as you touch your nose or eyes, to find entrance into your body. Finally, it will "white raft" in your bloodstream or lymph stream to find its favored host cells. These cells will have receptors on their surfaces that act as locks for the protein keys on the virus' surface. Once the virus has located the cells that have matching receptors, then, like a key fitting into a lock, the virus will open the outer coating of the cell wall and gain entrance. Once inside a cell, the virus becomes alive, takes charge of the cellular command center, and exerts control of the DNA in the cell nucleus, in order to convert the cell into a manufacturing plant for the production of more viruses.

The strand-like material inside of the virus is a single strand of nucleic acid, or RNA, which then commands the DNA, a double strand of nucleic acid (the double helix), to begin manufacturing the virus according to the virus' RNA blueprint. Thus, the virus begins to multiply and fill the cell. Soon, the attacked cell becomes so stuffed with invading virus that the cell wall will burst and release the newly manufactured virus. They will pour out into the host body to infect other cells. Soon, a multitude of cells are infected, swamping the host with the assaulting virus. And, thus, a disease!

Different viruses attack different parts of your body. Some viruses, such as cold viruses or flu viruses, target cells in the nose and throat and even lung tissue, while others, such as the polio virus, attack cells in the nervous system. By altering the DNA of your cells, the virus turns them into virus manufacturing plants in order to make thousands of more virus progeny. The relatively smaller size of the virus allows thousands to find a welcome (or unwelcome) home inside of a single cell.

Because of the way viruses invade cells, a disease caused by a virus usually develops fast. Within a seven-hour period, a virus can manufacture itself from one virus into 10,000 viruses. This rapid development is what makes antibody memory so important. If antibodies are quickly called into action upon the first sighting of a viral attack, your immune system can eliminate the invading virus before they have a chance to multiply. Once the viruses have been given a chance to gain entrance to enough cells to get a head start on their multiplication, it will usually take many days for your immune system to generate enough B-cells to turn into plasma cells and manufacture enough antibodies to eventually defeat the viruses.

Chicken pox, measles, and mumps are three childhood diseases caused by viruses. These diseases are highly contagious. They spread from child to child through touch or through the air. Most kids get vaccinations to protect them from these diseases. This immune protection occurs when the matching antibodies are already in the immune memory and can race to the scene the moment they get a report of the arrival of said virus.

A wide range of infections are called "cold." The common cold is a variety of different viruses that often generate an opportunity for bacterial infections of the throat and chest associated with a "cold." While your immune system is using energy to fight off the cold-causing virus, it is often easy for bacteria to get a foothold. Because many different viruses create the viral infection we know as a cold, and because companion bacterial infections are different, no vaccine has yet been developed to give protection from the "common cold."

Viruses have also been responsible for many human epidemics, such as the human immunodeficiency virus (HIV), Ebola, Zika, smallpox, and hepatitis, and, of course, the 1918 Swine Flu (Spanish Flu) pandemic.

As this book is being prepared, we are in the middle of the twenty-first century's global pandemic, COVID-19, a new and dangerous coronavirus. It is one of the many virus diseases that are referred to as zoonotic diseases, or "zoonoses." Where do they come from? Likely, the zoo part of this term provides a clue. The word applies to diseases that are transmitted from animals to humans, from infected animals, usually wild animals (animals from land, air, and sea). As for COVID-19, the source was bats, or most likely Chinese laboratory working on zoonoses to enhance the potency of the virus. The mysterious characteristics that make COVID-19 so different from other corona type viruses is that it is so highly contagious, spreading from not only direct physical contact, human-to-human, but surviving on doorknobs and other surfaces, to be transmitted from hand to face by unwary humans, while the virus is also airborne like measles and mumps, and it can be lethal to susceptible humans. The main difference appears to be the unusual arrangement of the amino acids in the protein spikes on the surface of the virus, spikes that are the connecting keys for entry into a human cell. This arrangement is a variation from all previously known coronaviruses.

Bacteria

Bacteria, like viruses, are one of the categories of germs, or pathogens, that are body invaders. They are the *wee animalcules* that Antonie van Leeuwenhoek discovered in his crude microscope, centuries ago. Unlike viruses, however, there are a vast number of bacteria that provide tremendous benefits.

Most bacteria are single-celled organisms that are encased in a hard, protective coat. They are shaped like balls, rods, or coils. They move through fluids by means of tiny, hair like fingers known as flagella. There are thousands and thousands of different bacteria. The bacteria that we are focusing on here are the ones that cause diseases, such as tuberculosis, typhus, typhoid, cholera, Lyme disease, whooping cough, diphtheria, and plague. Food poisoning is caused by the salmonella bacteria. But, while our focus is on the bad guys, there are thousands more bacteria that we need to sustain life, such as those living in your intestines that aid in the digestion. There are other good-guy bacteria that are required in dairies for the development of cheese and yogurt, or in wineries for the fermentation of grape juice into wine, or the bacteria in the soil of farms and gardens that aid in decomposing plants to return nutrients into the soil.

Bacteria appears to be the earliest of life forms, as anthropologists have found bacteria fossils dating back over 4.5 million years. Bacteria, like viruses, enter the body through natural openings such as the mouth or nose, as well as through cuts and burns.

While a virus will kill your body's cells by invading them and hijacking their DNA to produce more viruses, bacteria will kill your cells by releasing toxins—poisons—as they consume the nutrients intended to feed your cells. As your blood vessels roam around your stomach and intestines, your red blood cells pick up the food products—proteins, carbohydrates, fats, vitamins, and minerals—for delivery to each cell. Red blood cells are like the pizza delivery guy of your body. When the bacteria arrive, they intercept the delivery guys, take the food from them, and start gorging themselves. In the process, they create waste. The waste is poisonous to the surrounding body cells. Also, the presence of the bacteria and their toxic waste produces a local immune response, which causes fluid to accumulate around the cells—inflammation that squeezes the cells and damages them. The result is cell death—disease.

As mentioned, bacteria cause a wide range of diseases, including pneumonia, whooping cough, tuberculosis, typhoid fever, and tetanus.

Bacteria may attack specific body organs or move throughout your entire body. Each type of bacteria releases a different poison, resulting in different diseases. These infections can be treated with medicines called "antibiotics." Modern antibiotics focus on weakening the cell walls of the bacteria, or preventing the bacteria from being able to build its cell walls. As a result, the wall will tear, or burst, spilling its interior into the ravaging mercies of the white blood cells.

Penicillin is a familiar antibiotic. Penicillin works by fooling bacteria. Penicillin molecules mimic the shape and form of certain parts of protein molecules, which are the food for hungry bacteria. While bacteria can be dangerous, nobody ever said they were smart. The bacteria mistake the penicillin for their protein dinner, absorbing the penicillin molecules until they are full. There's no nutritional benefit for bacteria, because penicillin is the wrong molecule. The bacteria get too full to gobble up the protein they need to survive; technically the molecular material they need to maintain the protective cell walls. In their starving weakness, their cell walls collapse, spilling the interior contents, thus killing the bacterium.

Sulfonamides and other antibodies, such as tetracycline, also provide antibiotic weapons to defeat bacteria from being able to manufacture those proteins unique to bacteria and needed to build, or maintain, the walls of their cells.

Parasites

Some parasites are one-celled invaders similar to bacteria, examples of which are those causing diarrhea, stomach upset, and malaria. Others, such as worms, are multi-celled creatures. But while a bacterium is encased in a single rigid shell, parasites have soft shells. Because of this, parasites can move around more easily inside the human body. Parasites damage the body in many ways – they compete for food, block blood and lymph vessels, and cause obstructions, and they can produce toxins. Additionally, many parasites damage tissues and organs as they attempt to ensure that their eggs, or offspring, are available for reinfection of another host. This is usually done by making sure that their eggs are ejected from the body via waste, or the parasite offspring are in the blood stream for collection, from biting and sucking insects responsible for dispersing the parasite. These insects are called "vectors."

Parasites come in many shapes and sizes and cause many different kinds of diseases. Single-cell parasites are called "protozoa," and larger parasites, such as tapeworms, are animal parasites, while fungi and yeasts are parasitic plants.

Your immune system has a rough time dealing with parasites. If a malaria-carrying mosquito makes you a victim, it will bite your skin in an attempt to suck your blood, but during this procedure it will inject saliva containing the malaria parasite into your bloodstream. Normally, your white blood cells and antibodies would attack and destroy the unwelcomed invader, but the protozoa will quickly invade your red blood cells and hide inside of their protective shell. Eventually, the parasite will take up residence in your liver cells, where it is further protected from immune attack. So, the white blood cell search teams cannot find the malaria invader. Once the malaria parasite reaches a certain point in its lifecycle, it will destroy the invaded red blood cells and go back into the blood to be taken up by another mosquito. This destruction will reduce the number of red blood cells available to perform their tasks of carrying oxygen and food to the other parts of your body. As a result, you will become weak and, at times, feverish. Later, the parasites in the liver will emerge and infect new red blood cells, thus allowing the process to continue and making the parasite available for being sucked up by another mosquito and thus spreading the disease.

A trypanosome parasite, which causes a disease known as "sleeping sickness," enters your body from the bite of the tsetse fly, an insect found in some parts of Africa. This attack brings on the full host of immune system units, including massive antibody attacks. However, as soon as the B cells manufacture antibodies to attack the trypanosome, the parasite changes its molecular markers by allowing host proteins to coat the outside of its body. This fools the immune system into believing either that there is an entirely new invader, and a new battery of antibodies are manufactured to meet the new challenge, or that the parasite no longer exists. When a new attack is launched, the parasite again changes its coat. Then the trypanosome takes over, and the victim suffers the symptoms of sleeping sickness: Drowsiness, laziness, and long periods of inactivity. The victim's nervous system is slowly destroyed, and the infection eventually causes death.

Another parasite causes a disease called "schistosomiasis" and is a worm that evades immune attack by absorbing blood red markers onto its surface. This parasite infects either the liver and intestines or the bladder, where its eggs may easily be ejected into surrounding water during the body's waste removal processes. In the water, the eggs are ingested by snails, where eggs develop into new worms that, when expelled from the snails, will in turn burrow through the skin of swimmers. The worms lose some of the tail during their passage through the skin as an immune decoy to help the parasite evade immune attack. While the immune system is attacking the decoy, the tiny worms coat themselves with red blood cell antigens and hide in either the liver or the bladder. "Swimmer's itch," a common side effect of swimming in forest lakes, is caused by duck varieties of these worms, which while they cannot live in humans, they still leave worm proteins in the swimmer's skin, just enough to cause the irritating, swimmer's itch.

Parasites are more complex than viruses or bacteria. Doctors must use powerful chemicals to kill, or injure, parasites before the immune system can recognize them and launch an effective attack. Different chemicals are used for each type of parasite and sometimes a combination of drugs must be used to completely kill the parasite. Parasites are more common than people realize, and they are a constant danger to those who travel. Certain worms in both the U.S. and abroad come from vegetables that have contact with human waste and through meat found in poorly farmed animals, such as worms present in undercooked pigs and worms living in game, such as bears and deer.

Fungi

The fourth invaders are fungi. The fungi that cause disease in humans are parasitic plants that enter your body, or live on your skin, in the wet surfaces of your body openings, or under your nails. As plants, fungi have a shell exterior and are rigid and protective, like tree bark or the stem of a flower. The fungal invaders are usually single-celled and cylinder-shaped, but they grow into colonies of single-celled plants, their thin, threadlike tentacles weaving between cells and damaging the local tissue. Your immune system can do little to defeat fungi, because their shells do not have a protein marker on the surface to alert the white blood cells.

Athlete's foot, yeast infection, ringworm, thrush, and certain lung diseases are caused by fungi. Various bacteria in your skin help destroy fungi. In addition, medications called "fungicides" and some antibiotics can destroy fungal invaders. After fungicides damage the shell of the fungus, the immune system can then attack and destroy the interior of the fungus, causing reduced fungus growth and eventually the death of the fungus.

Foreign Cells

Normal human cells can also become foreign invaders when they are introduced into another person's body. This can be caused through blood transfusions, transplanted organs or cells such as bone marrow transplants, or the breakdown of a mother's placenta during pregnancy. Nowadays, blood transfusions are relatively safe thanks to the development of blood group testing prior to transfusion. This process matches the "r" markers on the surface of the red blood cells between the donor and the recipient and is commonly known as the ABO matching system. Giving a person unmatched blood will result in the person's immune system reacting to the foreign cells and killing, or rejecting, them. Today, such reactions occur when rare markers are present on the transfused cells, or when a sub-type of the blood group A is present, like A positive or A negative.

When an organ, such as a kidney or liver, is transplanted from one person to another, the molecular markers on the surface of the transplanted organ are seen as enemy invaders by the immune system of the person receiving the organ. Since the markers identify the new organ as not belonging, killer T cells and antibodies rush to the site, causing organ rejection and the subsequent death of the transplanted organ. This is why it is necessary to perform tissue typing to try to find organs closest to a molecular marker match as possible. This matching, when used with medicines that suppress the immune system, can allow a successful organ transplant. A classic example of the introduction of foreign cells into a person is a bone marrow transplant, in which immune cells are transplanted into a cancer patient so that they can fight the cancerous growth. In these cases, the person's original immune system is suppressed before the transplant, but often some cells remain active and fight the newly transplanted cells, causing a disease called "host versus graft."

Introduction of foreign cells can also occur when the placenta of a pregnant woman breaks down, and cells from her baby leak into her blood. If the father carries certain markers, such as the Rh factor or a different blood group marker, which is on the baby's red blood cells, then the mother's immune system will make antibodies that will seek out the infant's red blood cells and destroy them. This depletes the baby of oxygen-bearing red blood cells, causing a serious threat to its survival.

This condition can occur when the mother's blood type is Rh negative and the baby's blood has the opposite, Rh factor. This incompatibility can cause the mother's immune system's antibodies to attack, and destroy, the baby's red blood cells. When born with a low level of red blood cells, the baby will appear blue. This condition is treatable with good prenatal care and usually with immunoglobulin injections for the baby after birth.

Chemicals

Chemical invaders range from environmental agents to medicinal drugs, commercial dyes, and products. Basically, all chemicals are small molecules and, as themselves, can only weakly activate the immune system. Many chemicals can bind to large blood proteins where they can now alter the surface of these proteins and be attached by the immune system. When a chemical does this, it becomes known as a "hapten." Haptens can be medicines or they can be industrial chemicals, but whatever the source of the chemical, it usually induces an allergic reaction and causes tissue damage. Industrial chemicals such as formaldehyde can greatly affect the immune system, causing it to malfunction and lessen its effectiveness. Many airborne chemicals can bind to proteins and cells in the lungs and nasal passages, causing allergies, asthma, and tissue damage. This injury is usually the result of the immune system attacking the cells and killing them as innocent bystanders. Chemicals that become absorbed into the skin can cause allergies as a result of the immune system reacting against the invader. A classic example of this is poison ivy, where oils absorbed into the skin cause mild-to-severe, T cell reactions known as delayed hypersensitivity (we will talk more about this later in Chapter Eleven). Another example of contact, or delayed hypersensitivity, is the hive-like reactions when the skin comes into contact with objects containing nickel. This reaction often leaves a red imprint of the outline, or shape, of the nickel-containing object, such as a belt buckle.

Industrial chemicals, such as dyes and pesticides, often cause suppression of the immune system and leave the person defenseless against viral, and bacterial, infection. A good example is formaldehyde, a chemical sometimes used as a preservative or in the process of dry cleaning. This chemical can be absorbed either as a vapor or by direct contact with the skin. Either route, the effect is usually a loss of immune function and/or efficiency.

Chapter Eight
Vaccines: Manipulating the Immune System

Vaccinations are the reinforcements for your immune system army. They are the medical miracles by which your immune system can do a better job. Vaccinations give your immune system a boost so that your body can fight a wide range of diseases. Before there was a wide use of vaccinations, deadly diseases waged havoc throughout the world.

The Start with Smallpox

The story of vaccinations began with smallpox. Throughout the Middle Ages (A.D. 500 to 1500), smallpox epidemics swept through Asia, Africa, and Europe, killing thousands upon thousands of people. One out of every three children were likely to die from smallpox. European explorers carried this terrible disease to the natives in all of the American continents.

A victim of smallpox would first suffer intense pain, high fever, and chills, followed by an angry rash of hundreds of pimples, or "pox." These poxes would fill with pus before forming a covering scab. Hands and feet would begin to swell while the virus attacked internal organs and peeled away layers of skin. Many victims died. Those who survived were permanently pockmarked.

Smallpox was caused by a virus that spread from person to person through contact with the fluid from pox, or droplets in the air. But at that time, no one knew anything about bacteria or viruses; only that smallpox was a deadly disease for which there was no cure, and no protection.

In the spring of 1796, an English doctor, Edward Jenner, made a remarkable observation. He noticed that although smallpox epidemics spread through towns and villages, farm children who lived around cows and who caught a mild disease known as cowpox, never caught smallpox. These children seemed to possess some magical protection – they were immune to smallpox.

Jenner also noticed that country girls, who got cowpox from milking cows, often had pockmarks on their hands where the cowpox blisters had been. These pockmarks looked like the pockmarks on the faces of smallpox victims who had survived. From this observation, he reasoned that the cause of the disease was connected to the pus in the pox blisters.

Jenner also heard of a technique used in Asia. A smallpox blister was scratched with a sharpened stick, and then the same stick was used to scratch the skin of a healthy person. This produced a "small" case of smallpox – the person got the disease but did not die from it. Moreover, the person was prevented from getting smallpox again.

One day, Jenner performed a dangerous experiment. He found a country girl who had cowpox and punctured a watery blister on her wrist with a needle. Then, using the same needle, he scratched the cowpox fluid into the skin of a healthy boy, James Phipps. Several weeks later, Jenner injected James with the watery pus from one of his patients who had smallpox. Amazingly, the boy did not catch the disease. Jenner tried again to infect James Phipps with smallpox but found James to be immune.

Jenner was criticized by other doctors for putting James Phipps's life in danger. But the doctors had to admit that Jenner's experiment was a success.

Jenner called his new procedure *vaccinia*, from the Latin word for cow – *vacca*. News of the success of vaccinia spread.

So successful was the medical miracle born of the ingenuity of Dr. Jenner that by 1980, 184 years later, a program begun by the World Health Organization, led by a medical team headed by Dr. Donald Henderson, from the Center for Disease Control in Atlanta, Georgia, and involving thousands of workers around the world vaccinating millions, succeeded in completely eliminating the disease of smallpox from the face of the earth. Today, the only remaining samples of the smallpox virus are frozen, in secure vaults, at the CDC and at a similar center in Koltsovo, Russia.

The widening use of vaccinations as a method of preventing the spread of diseases was the work of French scientist Louis Pasteur. His studies of bacteria led to the development of methods to weaken microbes in the laboratory. He then injected the weakened microbes into animals to create protection against diseases such as anthrax in sheep and cholera in chickens. In 1881, Pasteur began work on a vaccine to prevent rabies, a fatal disease spread by the bite of a rabies-infected animal. Four years later, he had a chance to test his vaccine.

A boy named Joseph Meister was bitten by a rabid dog. His parents came to Pasteur and begged him to help. Pasteur had never tested the vaccine on a person, but he and Joseph's parents knew that without help, the boy would die. Happily, the vaccination was successful, and Joseph did not contract the deadly disease. By the early 1920s, vaccination was a familiar term.

How Vaccinations Work

Imagine that a classmate of yours has the mumps but doesn't know it and comes to school. Mumps is caused by a virus and is highly contagious. The mumps virus is spread by touching or is passed through the air in a cough or sneeze. If you hadn't before had the mumps yourself or a mumps vaccination, you would very likely catch the mumps. But if you had received the vaccination, you would not get the disease.

When a mumps virus enters your body, it is met by your white blood cells, particularly your macrophages, which are your frontline foot soldiers. Then, as you remember, they will send a message back to the command center where the helper T cells and suppressor T cells mobilize more troops to fight the invading mumps virus. When they call in the B cells to transform into plasma cells to manufacture antibodies, the war is on. The newly manufactured antibodies rush to attack and destroy the Mumps viruses. But, like many viruses, the mumps viruses multiply very fast, so you will suffer all the symptoms of mumps before the B cells can manufacture enough antibodies to defeat the viruses. You will have the mumps for several days before your body makes enough antibodies to overcome the virus. But once the mumps virus has been destroyed and the disease cured, the same antibody-producing cells will retire to the lymph nodes and become part of your immunological memory.

When a new mumps virus invades, your immune system can move into action immediately. Since the mumps antibodies are already in the B cell memory, the plasma cells can instantly produce new mumps antibodies to quickly overwhelm the invading mumps virus before it can multiply. Millions of antibodies for mumps destroy the invaders before there is any physical sign of the disease. You would not even know that a mumps virus had attacked you. That is immunity.

A vaccination artificially creates the same reaction. A vaccine injects a small amount of the virus that causes the disease with the intention of stimulating the production of antibodies designed for that disease, and then is stored in your immunological memory.

Through the twentieth century, vaccinations used two different approaches. One type used small amounts of the live viruses. The virus was weakened and altered so it would not cause a full-blown case of the disease but was still effective enough to produce a good antibody response. The second type of vaccination used killed viruses. A killed virus wouldn't cause disease, but the antigenic characteristics of even a dead virus created the antibody response necessary to produce immunity.

In 1954, Dr. Jonas E. Salk used killed polio viruses to produce the first vaccination against the dreaded disease. But, the killed virus produced a response that did not last long in immunological memory, so periodic booster shots were needed. Later, in 1961, Dr. Albert B. Sabin developed a weakened live virus vaccine that provided lifelong immunity. Even better, there was no needle. Instead, and happily for the children, just vaccine-soaked sugar cubes. The poliovirus being targeted is lodged primarily in the intestines. Today, because of vaccinations, polio has been eradicated in most countries.

Different vaccines have different periods of effectiveness. Some require a second, and sometimes a third, "booster shot." That's because the immune system reacts differently to different microorganisms and some vaccines are not as potent as others. Some invaders naturally produce a stronger antigenic response. The stronger the response, the more likely it is that your body will store their memory in your B cells, which, in turn, will trigger the needed production of antibodies for that virus or bacteria.

There are five different types of vaccinations: (1) killed pathogens, either viral or bacterial (2) live but weakened pathogens; (3) subunit vaccines; (4) toxoid vaccines, and (5) conjugate vaccines.

We have described (1) and (2), but not the other three. A subunit vaccine is one in which certain proteins are separated in the laboratory from the target pathogen in order to stimulate an antibody response without using the entire pathogen. These appear to offer fewer side effects, but are often not as effective. This technique has been given new emphasis with the exciting new COVID-19 vaccine program, as discussed below.

A conjugate vaccine is when both weak and strong pathogens are combined in order to get a strong antibody reaction. This technique has also been given new emphasis with the exciting new COVID-19 vaccine program.

A toxoid vaccine is like the diphtheria anti-toxin serum, which is made from the antibodies found in the blood of a horse that had first been vaccinated with the diphtheria vaccine: strange, but it works.

In today's laboratories, entirely new approaches to vaccinations are being explored; extracting DNA for pathogens to mimic an antigen, and getting an immune response, and from that response, immunity.

The Need and the Challenge

Many of the diseases prevented by vaccinations are the so-called childhood diseases, such as mumps, measles, and chickenpox. In children, these diseases are usually fairly mild. The same diseases in an adult are quite another thing. These infections attack organs not fully developed in children but that are, of course, in adults. For grown people, mumps is a serious disease. If a woman gets German measles when she is pregnant, she risks losing her baby.

In the past, epidemics of highly contagious diseases like mumps and measles commonly swept through schools. In recent years, these diseases are not as likely to occur during childhood because most children have had vaccinations. Thus, an unvaccinated child can reach adulthood without any protection against the childhood disease, not because of a natural immunity but because most of the children in that child's classroom have been vaccinated, and an entering virus would not have had enough potential victims to produce an epidemic. However, that unvaccinated child, upon reaching adulthood, can, unknowingly, be at serious risk.

In the early twentieth century, thousands of children suffered from diseases like whooping cough, polio, measles, chickenpox, and mumps. Diseases like measles and mumps were so wide-spread that most children would expect to contract both of these highly contagious diseases. In 1964–65, an epidemic of German measles (rubella) infected over 12 million Americans, killed 2,000 infants, and caused 11,000 women to miscarry their unborn babies. Now there are almost no outbreaks of measles, except in instances when anti-vaccine movements have convinced populations to refuse vaccinations. This has occurred in communities in Minnesota, Texas, Oregon, and New York, where there were groups refusing vaccinations with the unfortunate, but unnecessary, epidemics of measles returning with a vengeance. Although a vaccination can prevent measles, there is no cure for a victim who does catch measles. Because of vaccinations, measles is rarely found in the U.S.

However, an adult woman in Clallam County, Washington, died of measles in an outbreak among a community that had rejected vaccinations.

For instance, measles, which can become deadly, has been easily controlled in a population by a vaccine, which is highly effective and has rare side effects. The vaccine provided control. This has been important because measles is highly contagious, and can spread rapidly in an unvaccinated population. It is always a personal tragedy when a child suffers, or worse, dies, because of a disease which is easily, and safely, preventable. By the beginning of the twenty-first century, the United States had virtually eradicated measles. This is important because measles is one of the most extremely contagious diseases. The germs float in the air and can be infectious in a room days after the human carrier has been gone. But outbreaks of measles are back, largely due to social media groups inciting some communities to resist vaccinations. While the World Health Organization (WHO), working with the Global Alliance for Vaccination and Immunization, has been successful in approaching the total eradication of a range of infectious diseases such as polio, whooping cough, and other childhood diseases, the successes are being reversed. This is particularly and sadly, unfortunate because worldwide, over 100,000 children died in 2017 alone from measles and other infections, including pneumonia and diarrhea, which often follow.

Today, with more ill-informed groups around the world raising resistance to vaccinations, the WHO has listed vaccination challenges as one of the top ten threats to global good health. According to that organization, vaccinations are a process that, more than any other single worldwide action, has prevented two to three million deaths from deadly, but avoidable, diseases.

One of the common combined vaccinations given to children is the DPT shot. This provides protection against diphtheria, pertussis (German measles), and tetanus. Diphtheria is an often-deadly disease, involving a bacterium that produces a cell-killing toxin in a patient's throat, causing the throat to swell, literally choking the victims, usually children, to death. Before 1923 and the development of an anti-toxin vaccine, epidemics of diphtheria would sweep communities, with deadly consequences.

In the dead of winter in 1925, the remote Alaskan village of Nome was struck with an outbreak of diphtheria, and children began to die. The nearest source of the needed vaccine (an anti-toxin serum) was carried from Anchorage to a railroad terminal in Nenana, and from there it was handed to a pony-express type of non-stop relay between dog sled teams manned by brave men. In the middle of a major blizzard and freezing snow, these men handed the precious cargo from team to team as they rushed the vaccine, in record time, to Nome, along the now famous Iditarod Trail. This heroic exploit saved the children of Nome. This courageous act is still remembered today as the Iditarod Trail Sled Dog Race.

Although there are at least twenty different vaccines, and more being developed every day, some diseases defy vaccinations. Every year, the public is alerted to the need to get a flu shot. As was explained earlier, we cannot get one shot that will protect us for our whole lives because the flu viruses constantly change their outer coat – their molecular markers. For instance, a vaccine to protect against the Asian Type A flu virus will not protect against Hong Kong Type B flu. Each year, the prevailing virus is usually different from the type that spread the prior year.

Annually, public health officials must provide their best guesses about which flu virus will attack during the next flu season. Based on these guesses—educated guesses—a flu vaccine is manufactured. If the health officials guess correctly, many people will be protected. If they guess wrongly, the vaccine will not help, although, for the most part, a flu vaccine would likely reduce the severity of the flu, even if not exactly correct for that new strain of flu virus.

Vaccines for Parasites

As noted in our earlier chapter, many parasites constantly change their antigenic characteristics, which means that an antibody against the parasite will be ineffective after it changes its antigen. It is difficult, if not impossible, to make an effective vaccine against these parasites, such as malaria.

Malaria is treated with medicines such as chloroquine and quinacrine, but no vaccine has been developed to protect against malaria, a protozoan infection.

Another parasite that has defied vaccines is the single-celled trypanosome that attacks humans and some animals through the bite of the tsetse fly. The trypanosome keeps changing its appearance by shedding its outer coat. This fools the immune system into responding as if a new invader has arrived. A new attack is launched – and again, the parasite changes its coat. The process is repeated until the immune system is worn down. As you will recall from Chapter Seven, once the trypanosome takes over, the victim will suffer the symptoms of sleeping sickness – drowsiness, laziness, and long periods of inactivity. The victim's nervous system is slowly destroyed, and eventually the person dies.

Zika fever, yellow fever, dengue fever, West Nile virus, and others are also caused by the bite of a mosquito, but the mosquito is only the vector, and these diseases are viral in nature, so that vaccines can be fashioned for protection against them.

Are There Too Many Vaccines?

The question is often asked if infants are receiving too many vaccinations. Will an infant's immature immune system become overwhelmed by the injections of too many artificial pathogenic invaders? The question is a fair inquiry.

One hundred years ago, the only vaccination was for smallpox. Within the next fifty years, the number had increased to four: Diphtheria, pertussis (whooping cough), tetanus, and polio had been added to the smallpox vaccine. Beginning in the twenty-first century, children up to two years of age were receiving anywhere from eleven to twenty different vaccinations. The results have shown a significant decrease in the number of childhood diseases. And while this observation is encouraging, many parents have become concerned that too many vaccines might weaken, rather than strengthen, a child's body by overtaxing the immune system's capabilities to react properly.

Current studies in scientific and medical, peer-reviewed journals, multiple vaccines, even if given at the same time, introduce such a small number of antigenic proteins into your body, that multiple vaccinations have no adverse effect.

Vaccines that protect against a large range of bacterial and viral invaders are the real guardians of everyone's good health and do not overwhelm, or weaken, your immune system. Vaccines are an important aid in keeping your body in top running order.

And Then Came Covid-19

When the new, and potentially deadly, coronavirus, COVID-19, swept out of Wuhan, China to engulf the world in a new pandemic, the weapon to defeat this scourge was a vaccine. President Trump and Congress went on a "wartime basis," and Operation Warp Speed was initiated and funded with a nine-and-a-half-billion-dollar budget to develop, manufacture, and distribute a vaccine against this virus. Eight pharmaceutical companies undertook the challenge. There were different approaches. *Pfizer,* working with a German biotech company, was first to succeed with a RNA-based vaccine. *Moderna* was next, followed *by Johnson & Johnson*, working with cooperating research centers, who pursued the same approach. The process uses a biomolecule—a laboratory developed protein—using the genetic code of a protein to imitate the antigenic "spike" on the surface of the COVID-19 virus to trigger an antibody response to fight against the real virus, should it attack a human. And that, as we know, is what vaccines do. The new approach did not use the older technique of using the whole virus (killed or weakened) for the vaccine. Instead, the spike, or "s" antigen, is the protein marker, or badge, that identifies the COVID-19 virus and is the antigen that alerts the B-cells to manufacture antibodies to the attack. In this new approach, just the badge, or marker (the "spike" protein), is manufactured, inserted into fat bubbles called "liposomes" to hold its shape, and is what becomes the injected vaccine. No chance of getting the COVID-19 from the vaccination with only the enemy soldier's badge, and not the enemy soldier himself (or herself) in the vaccine.

Meanwhile, *AstraZeneca* pursued a vaccine system using one virus to deliver the "spike" protein into the patient, while *GlaxoSmith Kline* worked on developing a plant-derived synthetic "spike" protein, grown in leaves as developed by the Canadian, Quebec-based Medicago's spike protein. *Novavax* pursued a project incorporating pure COVID-19 proteins into nanoparticles as a vaccine. *Merck,* working with the Institute for Systems Biology and the research non-profit IAVI, sought to identify the molecular mechanism of COVID-19 and, using the technology in its miraculous Ebola Zaire Vaccine, to develop a COVID-19 vaccine. *CanSino Biologics* was working on a vaccine that combined an adenovirus with mRNA to deliver the "spike" protein to humans.

Applying the knowledge of herd immunity, it is believed that when at least eighty percent of the population has been inoculated, the virus will lose its ability to spread and ultimately fade from the populace. Herd immunity means that when a certain percentage of the community is immune, then the remaining people who have never had the disease or been vaccinated are, nevertheless, safe because there are not enough people left in the "herd" to allow the pathogen to circulate and multiply. The necessary percentage for herd immunity depends on how contagious the virus, or bacteria, is. Measles and COVID-19 are both highly contagious. Unfortunately, COVID-19, like the flu, continues to mutate requiring supplemental booster vaccinations.

Just as World War II accelerated the development of the aircraft industry and modern jets, Operation Warp Speed will have accelerated different vaccine techniques, so that, many other diseases will be candidates for these new vaccine techniques.

Chapter Nine
Cytokines: The Multitasking Messengers

Having already been introduced to the main units of your immune system army, we now need to examine how they communicate. There are several different, but coordinated, systems. We begin with "hormones," chemical messengers generated by your "endocrine" system. For example, your pancreas is an endocrine gland that produces such cell regulators as "insulin," the chemical that regulates glucose in cells throughout your body. There are other glands that produce messenger regulators such as "testosterone," "progesterone," and "gonadotropins." There are seven major glands in your endocrine system that produce hormones that are also chemical messengers, communicating random instructions to your immune system. As couriers, they travel like vessels in the rivers, creeks, and streams of your body—the blood and lymph flows—and deal with what we earlier referred to as innate immunity. In a sense, they are the Paul Reveres of the immune community, shouting-out, "The germs are coming; call out your killer cells and I'll give you a location." This we may refer to as communication for the innate immunity or frontline system, which we discussed in Chapter Four.

When the Revere's galloping horses are too slow, the body turns to the telegraph lines, the "voltage-gated ion channels" of the immune system, to carry messages at almost instantaneous speeds by electrical flows throughout the body. This system affects the whole immune system but particularly the adaptive immune responses against specific pathogens, which, as you remember, triggers cytotoxic T cells and their companion B cells that, transform into antibody manufacturing cells that in turn, launch their antibodies to destroy an invader. Rapid response times are sometimes an absolute necessity for the war against a virus.

Every cell in your body is cushioned like a leather helmeted cap around your head. This cushion is the cell membrane and consists of a double-layered padding called "lipids" (fat), which can become electrically charged. These molecules are charged through small electrical charges called "ions." These ions are special because they can generate electrical signals. When called to duty, these electrically charged ions push themselves through the fatty padding of the cell walls and create a tunnel known as an "ion channel." They do not use the Morse Code of our telegraph system, but they do get their message by electrical impulses because, by a series of inter-cellular-connections, these ions do become an electrical telegraph system for the immune army, sending communications to the various white blood cells of the immune system.

This communication system provides the speed and needed urgency in getting the right white blood cell army units to the scene of the immunological battle with the invading germ. While these ion channels provide electrical signals for other units (heart, lungs, stomach, etc.—you get the drift) of your body's operating systems, they are also an important signaling process for the immune system, working together with all of the other systems and the multitasking cytokines.

The Network

Now, as we dig deeper into our book of knowledge to explain the way your immune army protects you (or hurts you), we find an even more complex, more sophisticated, communication network. None of the message-carrying systems replaces the others. Instead, they all work together because the job of protecting all the organs of your body from all of the invading enemies requires a continuous, and coordinated, effort by all of the troops. So, while the dual systems of hormones and ion channels are flashing signals to your white blood cells in an easily understood, and seemly logical, order, your immune army also has an even more complex and remarkable component, the "cytokines."

A cytokine is a nanoscopic protein secreted by, sent-out by, or in official science-ese, "expressed by" many different cells, including white blood cells. Every white blood cell, as part of their other wondrous abilities, can also secrete (send out or send forth) specific cytokines. While it might seem logical to identify a certain cytokine with a very specific white blood cell, sometimes two different white blood cells might express the same identical type of cytokine. On the other hand, sometimes a white blood cell will vary the type of cytokines it produces. Sometimes a cytokine will refold itself and become chameleon-like, a different cytokine, taking advantage of the fact that the molecular configuration of each cytokine defines its function.

Cytokines are constructed by small polypeptides, forming tiny proteins that are then released from white blood cells for a wide range of services. If we were to assign a military duty to describe the cytokines, we would have to admit that the most appropriate description would be a motorcycle messenger and sometimes a messenger with weapons. First and foremost, they are produced by the cells receiving messages from victim cells and taking these messages to the appropriate T-cells to call out the troops. As we will explain more in Chapter Twelve, a polypeptide is a chain of amino acid molecules, while a protein is a larger molecule, in which a polypeptide chain is folded and twisted. It is the makeup of the polypeptide chain, and the way it is folded and twisted, that determines its function. At times, cytokines, as protein molecules, will, when necessary, change their own protein folds to change their functions – mysterious and magical. Given the need, a cytokine may even give a command to a cell of your body and change the nature of that cell. Put another way, every cytokine is a protein, but when nature refolds a protein, or the protein refolds itself, it becomes a different protein.

Operating in the extremely complex, interactive, web of immune system communications, cytokines play a major role in all transmissions and commands. This may be a little more information than you need in order to understand the role of cytokines in your immune system, but if a cytokine carries a message to cells nearby, or locally, it is called a "paracine action." If the message is sent to a distant cell, it is called an "endocrine action." Finally, if its message, or command function, is directed back to the cell that secreted it, it is called an "autocrine action."

Based on what is known today, there are hundreds of different cytokines performing their extraordinary tasks, but science has not yet found the complete battle plans of your immune system: There is more, much more, to be discovered. Indeed, even though more than 250 molecules with cytokine properties have been discovered, scientists have barely scratched the surface of who these immunological elves really are and what they can do.

As remarkable as the cytokines are, the understanding of their functions was not recognized until scientists in the middle of the twentieth century began to understand and appreciate the critical role played by these marvelous proteins. Our current knowledge began when Dr. Alick Isaacs, a virologist at the Medical Research Council (MRC) in England, began studying certain viruses that in laboratory tests stopped or "interfered" with the spread, or duplication, of infectious viruses. At that time, it was already known that antibiotics that were effective against bacteria had no effect on viral diseases. The major viral disease Dr. Isaacs studied was influenza, the flu. It had been observed by others that, in laboratory cultures, there were viruses that blocked the growth of other viruses, but no one understood the mechanism by which one virus could interfere with the spread of an infectious virus. While working on this research, he was joined by Dr. Jean Lindermann, a Swiss virologist. Together, they spent the next decade trying to unravel the mystery of this viral interference as a mechanism to battle viral diseases. In early 1958, Dr. Lindermann coined the name "interferon," as the molecule that could interfere with, and stop, the multiplication of infectious viruses. This was recognized in an article Dr. Isaacs published with Dr. Derek Burke, declaring:

"...Interferon is the name which has been given to a new substance which prevents the growth of a number of viruses without apparently causing any gross damage to the surrounding cells."

Pioneering work by Dr. Charles Dinerello and his colleagues at NIH in 1974 demonstrated the role of a leukocyte-derived "pyrogen" (any fever-inducing agent like a bacterium, virus, fungus, etc.), later to become "interleukin-1" (IL-1) causing inflammation and disease. Dr. Dinarello is considered one of the founding fathers of cytokines, having purified and cloned interleukin-1. This important step established the validation of cytokines as mediators of disease, particularly of inflammation. The cytokine IL-1 can induce fever, sleep, anorexia, muscle wasting, lymphocyte activation, and the hematological and hemodynamic changes typical of septic shock syndrome. This was a beginning. As knowledge of interferon expanded, scientists began to identify other types of interferon by the type of white blood cells from which they were produced. For example, those from the white blood cells, leukocytes, were called "interleukins;" those from white blood cells, macrophages, were called "interferons;" while other cytokines were being identified as being secreted from a variety of immune, and non-immune, cells. Of course, as more was learned, it became clear that all of the various interferon designations were still, really, sub-sets of cytokines. The identifying practice continued, and now a system has been established to identify all cytokines by the cells that produce them and their function, followed by numbers.

CYTOKINE SYMBOLS AND THEIR FUNCTION	
Inflammation	IL-1, IL-6, IL-8, IFN , TNF
Cell-mediated immunity	IL-2, IL-12, IL-15, IFN ,
Humoral immunity	IL-4, IL-5, IL-10, IL-13, TNF
Hematopoiesis	IL-3, IL-7, IL-9, IL-11
Allergy	IL-4, IL-5

With permission of Dr. Terry M. Phillips

The Dark Side

There is, unfortunately, also a dark side to cytokines. There is evidence that over-enthusiastic cytokines are involved in the autoimmune disorders, which we will discuss in Chapter Ten. As research into the world of cytokines and the development of more knowledge of the causes of autoimmune diseases, the role of the dark side of cytokines increases and soon should be understood and fully revealed. It is already known that cytokines can have not only anti-inflammatory but also have pro-inflammatory effects. As magical as these little molecules may be for our protection, we must remember that they are not possessed of independent, active intelligence, but rather are bound to the rules of nature. When there is a dysfunction in your body, the cytokines can become misdirected and misread the "Hey I belong here" badges ("receptors") on your own cells and activate an attack against the cells that are "self," and that means "trouble."

And then, there are "cytokine storms." An invading pathogen, like the avian flu virus, which is highly infectious and vicious, can create a rushed and chaotic immunological response. The army is caught by surprise by the viciousness of the new enemy and responds with a major call to arms: This is the Pearl Harbor of the immune system. "This is all out war!" This communication, in some instances, can cause a sudden and overwhelming release of cytokines. Uncontrolled regiments of white blood cells rush to the attack in great disorder, pouring more white blood cells than needed to the identified site. The result on the sufferer will be a high fever, swelling and redness, extreme fatigue and nausea, and sometimes death.

One writer has described a cytokine storm as the event when a perfectly healthy immune system releases over 150 known inflammatory mediators. It has been postulated that the devastating swine flu epidemic of 1918 that killed over 500,000 young, seemingly healthy adults is now thought to be the result of a highly toxic flu virus that suddenly, and unfortunately, activated too many uncontrolled cytokines, and other pro-inflammatory proteins.

Reported cytokine storms arose in 2005 when thousands of immune systems went into high gear as a result of a pandemic attack of avian flu that, unfortunately, became deadly as cytokine storms struck vast populations. Cytokine storms have also been suspected in some of the violent reactions indicated by severe lung involvement leading to death in the COVID-19 pandemic. There are surely others.

"Immunomodulating agents" are chemicals, or molecules, that can interact with the cells of the immune system and control their activities. Both hormones and cytokines fall into this category. As communicators, they are both involved in "endocrine, paracrine," and autocrine "signaling." Nice, important words, but what do they mean? Endocrine signaling is performed by the chemicals referred to as hormones flowing throughout your body, although the term is also used to describe cytokines that act on non-immune parts of your body. For example, the hormone "insulin" affects the sugar used by cells in your body. In the same way, the cytokine interleukin-1 can interact with the "hypothalamus" in the brain to induce sleep and raise body temperature.

Paracrine signaling involves the flow of either hormones or cytokines released to interact with cells in the local vicinity. For example, the "pituitary gland," which lies under the brain, produces the thyroid stimulating hormone, which controls the hormone production by the thyroid and the interaction between gamma interferon produced by T cells and macrophages, which, in turn, causes the macrophages to become more aggressive. Both are paracrine interactions.

Autocrine signaling is when the cell sends a signal that reacts with itself and is more common in cytokine interactions. An interaction means a cytokine can actually have a "cell changing effect" on a white blood cell, for example, the cytokine interleukin-2 is produced by activated T cells and that cytokine can interact with interleukin-2 receptors on the original T cell and stimulate it to multiply itself. This is called "clonal expansion," and is an example of the multiple, or joint, responses of the immune system in its role, which is to protect your body.

Chapter Ten
When It Fails and When It Attacks: The Nightmare

We have progressed through these pages with the necessity of assuming for our discussions that everyone's immune systems are the same. Unfortunately, they are not. Most are, but some have problems. So, to begin this discussion, we ask four questions, which we will also answer. Is it possible for your immune system to stop working? Or to overwork? Or never to have worked at all? Or become traitorous against you? The answer is "yes" to all four questions.

Your immune system can be defective in whole or in part at birth, or it can break down later. If your immune system is defective at birth, that is, if you are born without a functioning immune system, the disorder is called "immunodeficiency." This means that you are born without T cells or B cells. This disorder is specifically called "severe combined immunodeficiency, or SCID." There is also an "ADA deficiency SCID (Adenosine deaminase deficiency)" in which you are born with very low levels of white blood cells. This low level of immune fighting units is often not recognized until later on in childhood, or into adolescence, sometimes even into adulthood. This occurs if you are born without fully developed stem cells, the essential generating cells that produce and supply white blood cells. Without enough white blood cells, your body is not equipped to give you effective immune protection. This is known by a number of different names, as we shall discuss.

On the other hand, when your immune system becomes overactive, it is called an "allergy," and finally, when your immune system turns against you, this is called "autoimmunity." (see the next chapter).

A child born without any immune system, no T cells or B cells, SCID, will unfortunately be quickly overwhelmed by infections of all sorts from the natural world of germs in which nature is awash. That child will have no chance of survival without medical help. Such a child is referred to as suffering from the "bubble boy disease." This unfortunate, and unwelcome, honor was given to David Vetter, a child born in Houston, Texas, in 1971. He was nicknamed "the bubble boy" because when it was realized that he had no functional immune system, he was quickly put into a totally germ-free enclosure. At first, he was contained in a crib, or playpen type of enclosure, totally isolated from any exposure to the thousands of microbes that are in our normal living spaces. He was cared for through long, sealed gloves built into the unit, similar to the type of apparatus used in level four research of dangerous microbes at the Center for Disease Control (CDC), the National Institutes of Health (NIH), or at similar research laboratories. As David outgrew the enclosures, the National Aeronautics and Space Administration (NASA) built a small space suit for him so he could spend time in the backyard with his family. As he grew, it was understood that he would never be able to leave his protected enclosures unless he could receive a new immune system through a bone marrow transplant of white blood cell producing stem cells. It was determined that if he could receive and his body would accept, without rejection, a bone marrow transplant, new stem cells would begin to manufacture white blood cells and give him a chance for a normal, healthy life. Some bone marrow transplants had been successful with ADA deficiency SCID patients, those who only needed a boost for their existing immune systems. However, David needed an entirely new set of stem cells.

When David was twelve years old, the family and the doctors decided to attempt a bone marrow transplant from his sister, who had a healthy immune system and was a close immunological match. Unfortunately, and unknown to all, her system hosted a dormant, lurking virus. This virus, while harmless to his sister, was able to spring to life from the transplant and to overwhelm the immune system newly transferred to David. Neither his new stem cells could generate a new immune system, nor any other medical support treatment could stop the rapid onslaught of the attacking virus, and David died.

Not all SCID victims are that unlucky. The use of healthy stem cells given to a child through an intravenous catheter can cause new cells to travel into the bone marrow to begin making healthy white blood cells. Today, most SCID children have a better chance of survival if the transplantation is performed quickly, after an early diagnosis of SCID, and with the application of new advances in gene therapy.

If the defect in the immune system is the absence of B cells, then no antibodies can be produced. This disorder is called "Hypogammaglobulin Anemia" and is treated by injections of gamma globulin. "Gamma globulin" is the part of the blood that contains the antibodies produced by B cells. The gamma globulin provides the patient with a temporary dose of antibodies. With injections every other week, the patient can lead a reasonably normal life if weekly injections can be considered normal.

If the defect in the immune system is imperfect white blood cells caused by a missing chromosome at birth, the unfortunate result is congenital heart disease, hypocalcemia, and retarded development, among others. The disorder is called "DiGeorge Syndrome." There is no known cure, but there are a wide range of treatments depending on how the patient's body reacts to the defective white blood cells.

Like an army with many specialized squads and companies, each of your immune units has its own specialized function(s). If there is a defect in anyone, or combination of these units, then there is a corresponding disease. Genetic immunodeficiency includes a list of approximately seventy defined disorders, but three—SCID, Hypogammaglobulin Anemia, and DiGeorge Syndrome—are the most common.

New Approaches

In 1972, thirty-seven-year-old Stanley N. Cohen, an American geneticist, and thirty-six-year-old Herbert Boyer, were the first scientists to transplant genes from one living organism to another. This discovery was the beginning of one of the most exciting developments in biomedicine and biotechnology, otherwise known as gene replacement, or "gene therapy." While this discovery has had an impact on hundreds of areas of science and medicine, it was the beginning of a real cure for SCID and its related disorders.

Gene therapy is the art of using "enzymes" to remove and transplant specific genes to modify a defective gene or insert a missing gene in order to change the DNA controlling a failed organ or system.

Four decades following the work of Cohen and Boyer, Colton Ainslie, born prematurely, was diagnosed with SCID. Horrified, his mother, Jessica, contacted Dr. Donald Kohn, a leading specialist in ADA-deficiency SCID patients. Dr. Kohan had been working with other researchers in the new field of gene therapy. At the hospital at UCLA in California, Colton was given a low dose of chemotherapy followed by the newly developed gene therapy. It worked. While Colton does require regular treatments, he is now leading a relatively normal life. Other successes have followed, including Colton's baby sister, Abbygail, who was also born with ADA-deficiency-SCID, but luckily was rushed to Dr. Kohn's care for early treatment, and has been completely cured of the disorder.

Gene therapy is now used by more and more doctors to treat SCID and other immune deficiencies. Bone marrow cells are removed from the patient, and a corrected copy of the defective gene is inserted into the cells, which are then returned to the patient's bone marrow to refresh their stem cells. If all goes well, a normal immune system will result. The earlier this treatment is done, the greater the likelihood of success.

Your body parts, whether hair, heart, lungs, skin, bones—you get the idea—all are made up of cells. Inside each of your over thirty-seven-trillion cells is a nucleus, and inside each and every nucleus, there are "chromosomes." Each chromosome is a single, tightly coiled molecule (which if stretched into a straight line would probably be about two inches long). Two of these single molecules coiled around protein molecules, that are twisted together in a ladder-like formation called a "double helix," contain your "DNA" – short for "deoxyribonucleic acid."

You will have inherited twenty-three chromosomes from your father and another twenty-three from your mother, and these two strands of DNA, the combined forty-six chromosomes, contain your genes (hundreds of thousands of genes), which define who you are, what you look like and for the purposes of this book, how good your immune system is going to work. Your DNA is the personal encyclopedia of you. Your genes are the blueprint for every detail of your body and all its parts, creating a body that is uniquely yours. It is the absence of an essential gene, or a defect in a gene, that produces the individual with SCID. For this reason, the concept of providing the missing gene, or replacing a defective gene, makes gene therapy an exciting approach to cure not only SCID but for many other disorders, or diseases, that are genetic in nature, such as cystic fibrosis, Tay-Sachs disease, and down syndrome. This means that somewhere in the DNA of your ancestors, a gene was "out of whack," or was waiting to be mated to a partner with a certain genetic predisposition, like an army that recruits only goofballs and misfits. You could probably avoid any of these missing, or defective, gene problems if you could pre-test and select your ancestors, but that opportunity awaits the fuller development of the science of designer babies or, of course, time travel.

Then There's Aids

While SCID is a rare disease, or disorder, an equally deadly but more widespread disease is another form of immunodeficiency, "AIDS," which stands for "Acquired Immune Deficiency Syndrome." As its name suggests, this disease is not a result of a genetic defect at birth. It is caused by a virus that destroys the helper T-cell white blood cells, the important generals in the cellular immune army. This virus, "Human Immunodeficiency Virus (HIV)," causes a defect in your immune system, not a defect from birth, but it is a virus that the victim acquires.

The appearance, and invasion, of this virus in the world population has, over a period of almost forty years, created a grim and deadly epidemic. The disease is spread by the exchange of bodily fluids from sexual activities between partners, sharing of hypodermic needles between drug users, or careless medical practices during blood transfusions, in which the donor is carrying the HIV virus, or passing from an infected mother to her child in childbirth. Unfortunately, this terrible disease has claimed millions and millions of lives around the world, and also, unfortunately, the epidemic has not been brought under control, although studied by the World Health Organization (WHO), the Center for Disease Control (CDC), the National Institutes of Health (NIH), and health organizations in various countries. New medical tools do hold promise, but there has not yet been universal success.

What happens? The HIV virus invades the victim's body, then seeks a particular type of white blood cell, carrying a specific receptor, to use as a virus manufacturing plant. Remember, viruses cannot reproduce themselves. They need a host. So, the HIV virus attaches to a white blood helper cell, then penetrates the cell wall, and enters into the nucleus within the cell. Inside the cell, the virus sheds its protein coat, empties its own nucleic acid into the host cell, changes the DNA code within the cell, and finally causes the host cell, with its new DNA, to become a manufacturing plant for more viruses. Ultimately, the host cell dies, but not before millions of new, deadly viruses have been manufactured.

The HIV virus is especially dangerous because of the host cell it attacks – the helper T cell. As you will recall, the helper T cell is one of the two generals commanding, and controlling all the other units of your immune army. As these helper T cells are destroyed, all commands become weaker, and communications become garbled until no commands are made. When no commands are given, no new antibodies are produced, and no new killer T cells or B cells are produced. As a result, the invaded body then loses its "immunity." As a consequence, the victim of the HIV virus now has AIDS and is susceptible to attacks by other invaders – attacks by other viruses, bacteria, parasites, or fungi. These are called "opportunistic" infections. Lung infections, such as pneumonia and skin cancer, such as Kaposi's sarcoma, are prevalent, although these are only some of the illnesses that victims of AIDS may suffer.

While a bone marrow transplant may work to restore the immune system of SCID victims, a bone marrow transplant will not work with an AIDS victim because the HIV virus will destroy any new helper T cells as they develop. Worse, HIV is an extremely aggressive virus, reproducing much faster than other invading viruses, and even worse, the virus, when challenged, can easily mutate, changing its protein markings. Worst of all, when the virus invades white blood cells, it can remain inactive for a long time, even up to ten years. Some people have been found to house HIV viruses in their blood but not show any sign of the disease, AIDS. This makes the HIV carrier a walking disease-producing timebomb, which can explode on the victim spontaneously and can also carry the disease to other unsuspecting people. A blood test for HIV antibodies could provide the answer, but another test, called a "western blot," would have to be performed to check for the presence of the HIV antigens rather than the presence of anti-HIV antibodies. The western blot detects the protein markers shed by the virus, and not the patient's antibody response to the infection.

At this time, there is no known cure for AIDS, but three-drug anti-HIV antiretroviral treatments have been successfully used to slow the progress of the virus. This antiretroviral therapy is referred to as "ART" and is a cocktail of drugs, usually designed in each individual case by the treating physician because one formula does not fit all. There are also some medications being developed that have shown promise of offering limited protection against the virus. They are the hope for the future.

The Traitor – Autoimmunity

You will recall that one of the magical characteristics of your immune system is that your white blood cells, and all of their progeny, recognize the badges, or protein markers, that tell them that all the cells of your body are you. As we have already discussed, before birth, every cell in your body was given a molecular badge. Your immune system has all of your millions and millions of parts catalogued, and as they flow through your body, they keep a careful immunological eye on every cell to make sure that all are displaying their badges. But then, when your immune inspectors make an error and misread a molecular badge, mistaking normal self-cells as foreign, a command goes out: "There is a foreign invader in our midst!" Killer T cells and antibodies then launch an attack on your own body. This attack is an autoimmune response.

Unfortunately, there are quite a number of diseases that are autoimmune reactions. Some are organ-specific and some are more general. "Multiple sclerosis (MS)" is an example of an organ-specific autoimmune attack. In MS, the attack takes place within the brain and spinal cord. The immunological army attacks and eats holes in the protective myelin lining of the victim's nervous system. Myelin protecting the nerves can be compared to the plastic or rubber-like protection on your electrical cords. If the protective coating is removed, or torn, from an electric wire, there would be an electrical short to cut off, for example, your electric light or even cause a fire. This is the same thing that happens when the myelin sheath is removed or punctured. In multiple sclerosis, the attack is on the central nervous system. In other autoimmune diseases, the attack is on the peripheral nervous system – to the nerves controlling muscles, resulting in weakness or paralysis. This occurs in the muscle disease "myasthenia gravis," an autoimmune disease wherein autoantibodies block, and sometimes destroy, receptors at the neuromuscular junction, preventing nerve impulses from triggering muscle contractions. Another example is "Guillian-Barre syndrome," where antibodies directed against the nerve coat prevent electrical impulses from triggering nerve impulses, resulting in temporary partial paralysis. Autoimmune attacks can be focused on specific anatomical areas such as the linings of the joints in "rheumatoid arthritis," or the linings of the salivary glands, as in "Sjogrens syndrome," but when the attack is not against any specific organ, but generally throughout the entire body, it is referred to as "systemic," as in "systemic lupus erythematosus."

What causes the immune army to turn traitor in the form of an autoimmune disease is unknown. Genetics and the environment are two predisposing factors. The nature of certain genes and sometimes the mutation of certain genes causes the body to more readily accept an attack that will trigger an autoimmune reaction from an infection, either viral or bacterial.

In the winter of 1976, a young army recruit at Fort Dix in New Jersey fell ill with the flu and died. A blood sample revealed a flu virus described as H1N1, a virus identified by the CDC as Swine flu, similar to the 1918 Swine flu outbreak (sometimes called "the Spanish Flu"). The 1918 swine flu was one of the world's most serious flu pandemics that sickened over 500,000 people and killed an additional 50,000. The CDC, fearing another deadly national swine flu pandemic, called for a team to create a flu vaccine to meet the pending outbreak. A national vaccination program was developed, and nationwide vaccinations began in October of 1976. Shortly after the program began, doctors around the country began to report an upsurge in Guillian-Barre, an autoimmune disease that causes paralysis of the limbs, sometimes temporary, sometimes permanent, and sometimes deadly. Because of this unexpected side effect of the vaccine, the program was temporarily shut down, and new plans were made. However, over the following months, more than fifteen hundred cases of autoimmune disorders were identified as having been caused by the flu vaccination. In the end, there was no swine flu epidemic after all, and, in fact, no swine flu occurred in the United States during that year except a few isolated cases where the victims were quickly identified as farmers who had been in contact with infected pigs.

However, an autoimmune reaction to flu vaccines still remains as a remote, but very real, unwelcome potential. Since Guillian-Barre is also associated with some viral infections, the mechanism of how a rogue protein from the vaccine could cause the body to suddenly attack itself is still unknown, although a protein called "P2," derived from chicken egg protein (vaccines were made in eggs), is known to immunologically react with a similar protein found in human myelin.

For example, in 2016, when the "Zika" virus was rampant in South America, it caused a well-understood fear among pregnant women, who feared it would cause their babies to be malformed. This fear overshadowed even the fear of an also possible resulting illness, Guillian Barre Syndrome, which paralyzed some of the victims following a case of Zika.

Fortunately, some treatments, such as immunosuppressive drugs and the infusion of antibodies from transfusions of blood from healthy donors, have been effective in treating some of the autoimmune diseases. However, a complete cure has still not been found. There are at least twenty-two diseases, such as Type 1 diabetes, Hashimoto's thyroiditis, inflammatory bowel disease, pernicious anemia, encephalitis, rheumatoid arthritis, rheumatic fever, and SLE, that have been identified as autoimmune diseases, or, in simpler terms, a result of the body attacking itself.

Besides the usual suspects, evolving knowledge of the role of cytokines (discussed more in Chapter Nine) in the complex organization of the immune system is providing evidence that sometimes cytokines themselves may play an unintended role in autoimmune disorders.

Chapter Eleven
Allergic Reactions:
You Get the Itch

Unfortunately, not all immune systems are equal. At least two out of every ten, persons will suffer from some form of allergy, a disorder that arises from a very active, or "hyperactive," immune system. Among the thousands of invaders that are seeking to find a home in your body, there are many that are not among those we have identified in Chapter Five, but are irritants known as "allergens." Those among us who are affected by allergens are suffering from a sensitive reaction, a condition known as "hypersensitivity." Or more plainly stated, if you are a person with a hypersensitive immune system, then your immune system is overreacting to the allergenic invaders.

Unfortunately, frequent allergenic invaders are pollens – the almost invisible dust-like particles that are plant seeds floating in the air from a variety of trees, weeds, and grasses. For example, in the spring, as you inhale the sweet perfume of a flowering honeysuckle, you may also be inhaling the invisible pollen from a sycamore tree, from which you may suffer an allergic reaction. If you are sensitive to that pollen, you might start sneezing, your eyes could water, and your nose may run. Hay fever, a common allergy that causes many watery eyes and runny noses, has nothing to do with hay. The villains are those microscopic pollen grains from a wide range of plants. Because the start of the pollinating season coincides with the springtime mowing of hay, the allergy has been dubbed "hay fever."

There are many different allergenic invaders. Besides plant pollen, a glass of milk, a piece of cheese, a sesame seed roll, dust, feathers, and even dander (flakes of dried skin) from a dog or cat can trigger an allergic reaction. Almost anything can cause an adverse reaction, but some things are more likely to be allergenic—to cause allergies—than others.

If you are one of the twenty percent of those who suffer from some type of allergy—one of those people who are "hypersensitive"—your body is either over-producing a special, potent, series of antibodies, or has allergen-reactive T cells. An allergic reaction can be immediate or delayed, occurring immediately (right away) or hours later, or even days (delayed), after exposure to the allergen. For instance, a bee sting can cause a violent allergenic reaction almost immediately, while an allergic reaction to poison ivy, or poison oak, may be delayed and not appear until two or three days later. For some people with allergies, a bee sting or a shot of penicillin can cause an immediate and extremely serious reaction. Besides red, itching hives and shortness of breath, those with this allergy can suddenly be in a death struggle as they gasp for oxygen as their bronchial tubes begin to swell shut, blocking their air passages. This serious condition is called "anaphylaxis."

In an emergency an immediate injection of Epinephrine will be needed. Persons with known sensitivity will carry emergency injectors as prescribed by their doctors.

Some children and adults are allergic to cow's milk. (Many other people can't drink milk because of lactose intolerance, caused by the absence of an enzyme. This is not the same as an allergic reaction.). The connection between drinking milk and stomach pains and diarrhea has been observed for over two hundred years. Children have a greater chance of developing an allergy to cow's milk if they are bottle-fed rather than breast-fed. This is true for children of non-allergenic parents as well as children of allergic parents. A mother's milk provides a source of immunity for the baby, protecting her baby's developing intestines from the foreign protein in cow's milk until the infant's own immune system develops. This also explains why physicians recommend formulas and supplements rather than cow's milk during the child's first year.

Some allergies change over time. If you are an allergy sufferer, you may grow out of your allergies. Unfortunately, you may grow out of one allergy only to develop new ones.

How Allergies Work

If you are not hypersensitive, any allergens you come in contact with will be met and captured in the moist tissues of your throat, bronchial tubes, or lungs and dealt with by the first line of defense, the macrophages. These white blood cells, along with certain other types of white blood cells, encircle and destroy the invading allergens, and you will never realize you have received an unwelcome visitor. The defense system does not leave all the tasks to the infantry. It also calls up the rockets fired from the B-cell commandos, the antibodies. When an allergenic microorganism enters your body, the macrophages send the cytokine messengers to headquarters, where the T cell generals, helper cells, and suppressor cells signal the B cells to begin plasma cell antibody production. The particular antibody regiment designated as the first rockets fired, defending unit against the invasion of allergens, are the specifically designated antibodies, "immunoglobulin E," identified in the shorthand, "IgE," to rush to your defense. The IgEs circulate in the rivers and streams of the blood and lymph systems. When these antibodies find, and latch onto, an antigen, they capture and transport the allergen to a floating battle station, a "basophil," or a stationary battle station, a "mast cell." Once the allergen is connected to the battle station, the battle station releases a combination of destructive chemicals along with "histamine" to destroy the invader. While histamine has numerous functions in your body, with respect to allergens, it also acts on small blood vessels to make them permeable and leaky, an advantage for allowing other immune cells and antibodies to migrate into the site of the allergen location and destroy additional allergens.

In an immediate hypersensitive reaction, an allergen causes the command center to become confused. The B cells manufacture more IgE antibodies than are required—ten times more—and they, in turn, overload the basophils and mast cells. An excessive amount of histamine and other histamine-like chemicals are released. The unfortunate result is a tempest of chemicals – much more than needed. This release of excessive histamine and histamine-like chemicals become the unwanted body enemy. Large doses of these chemicals cause watery eyes, runny nose, itching, wheezing, sneezing, and even vomiting or stomachaches – all the symptoms of the allergic reaction that the overload has generated. With a dangerous allergic reaction like a bee sting, the overwhelming flood of histamine and histamine-like chemicals can cause the soft tissues found in the throat and bronchial tubes of the lungs to swell so rapidly that the sufferer cannot breathe. Histamine can also cause contraction of the muscles lining the blood vessels in the affected area, thus increasing the problem by lessening blood flow. This is the cause of "anaphylaxis," and, as mentioned before, it can be fatal.

Picture the mast cells, or basophils, as Velcro-covered balls that collect the delicate antibodies along with their allergen-pathogenic prisoners. As the IgEs and allergens load onto the ball, they set off small, but lethal histamine packages. If you are a person with an immediate hypersensitive predisposition, then your immune system does not control the flood of IgEs and the ball becomes overloaded and explodes with an overwhelming, and unwelcome, dose of histamine.

The same thing happens with food allergies. Food allergens enter the bloodstream through the walls of the small intestines, where they are met by roving IgE antibodies. Following their standing orders, they carry the allergens to the basophils or to the mast cells, and again, by reason of the confusion at headquarters, set off allergic reactions in your body.

Worse, if you have more than one allergy, your basophils and mast cells are handling an influx of food allergens but not causing an attack, and then, for example, you inhale a pollen for which you are allergic, then you might tip the scales, overload the basophils and mast cells, setting off the histamine explosion that is an allergic reaction.

An allergic reaction to food may occur no matter its form. If you are allergic to eggs, then you may have a reaction when you eat any food made with eggs – cakes, cookies, quiche, egg noodles. Some individuals even react to egg proteins in the atmosphere when they are in close proximity to a kitchen. This could become significant if you receive a vaccine, like a flu vaccine developed in eggs, then an allergic reaction to that offending allergen can trigger an unwelcome reaction. If your allergy is to corn, then corn meal, corn starch, corn syrup, and corn oil can all cause a reaction.

A delayed allergic reaction results from an entirely different set of immune system responses than that of an immediate allergic reaction that we have just described. Delayed hypersensitivity is not caused by over reactive IgE antibodies, but instead by an overstimulation of the T cells. Remember that white blood cells are programmed into various types, such as lymphocytes—helper T cells, suppressor T cells, killer T cells, B cells, and leukocytes—macrophages, neutrophils, basophils, and eosinophils. Delayed hypersensitivity occurs slowly, as the name suggests, usually over a period from one day to many days, sometimes even a week. The rash caused by poison ivy is a form of delayed hypersensitivity reaction. If you are predisposed to hypersensitivity, then when your skin touches the oil from a poison ivy leaf, sensitized T cells rush to that area of the skin to attack the allergenic oil. When they do so, but they not only destroy the invader, they also cause damage to the surrounding skin cells. As a result, the invading T cells cause pressure in the tissue and nerve endings, resulting in the release of inflammatory neuropeptides which cause red wheals and raised pink rashes to pop out on the surface of your skin. Because it may take the T cells a day or two to process the allergen, the reaction is delayed. You then suffer with an itchy rash for days after you walk through a patch of poison ivy. Other examples of delayed hypersensitivity are often known as "contact dermatitis." In such cases, the allergen is a chemical present in a common article that comes into contact with the patient's skin. Shoe dyes and minerals in antique jewelry are common examples. In the first case, the dye leaves an outline of red, itchy skin where the shoe leather has come into contact, and in the other case, minerals, such as copper and nickel, can come out of the metals after reacting with sweat. These cause itchy areas on the earlobes where earrings have been, or an itchy finger beneath a ring. In both cases, the allergen reacts with T cells, and localized tissue damage results from the T cells attacking the source of the allergen.

The rejection of a skin graft or an organ transplant is usually another form of delayed hypersensitivity, although in some cases, immediate hypersensitivity occurs. After an organ transplant, the surgeon may report that the operation was successful, but the doctor and patient both must wait to see if the patient's immune system will reject the transplanted skin or organ. This means they're waiting for the possibility of a delayed hypersensitivity reaction. Tissue typing, or matching the donor cells with the recipient's cells, helps to reduce the likelihood of rejection. Cells are studied to determine if there's a close molecular match between the cells of the organ recipient and those of the donor. In addition, medicine that slows down the immune system's response, called "immunosuppressive" drugs, are also used to help make the organ transplant successful.

Your physical condition can also affect an allergic reaction. If you are cold, upset, or ill, you may be more susceptible to a white blood cell, or antibody overload, and suffer a hypersensitive reaction, as the signals to your brain, affected by your condition, can affect the immune system commanders.

Preventing Allergies

The best way to protect yourself from an allergic reaction is to know what you are allergic to and avoid exposure to that allergen. In order to know what substances, you are allergic to, it is necessary to see an allergist, who will test you against different known allergies. This can be done by performing a patch test. Here, the allergist places small patches of gauze soaked in an "allogen" solution on your arm or back, then a day or so later, reads the reactions. Each different allogen is placed on a different patch. Allogens can also be injected into the skin in a specific pattern, and the same procedure is used to look for reactions. This is called a "prick test." Although many allergists still use the patch, or prick test, a chemical test is available which measures the amount of antibodies that are binding to allogens attached to paper disks. This test only requires patient serum and is called the "ELISA" allergen test. The ELISA test is good for detecting antibody-mediated reactions, while the patch and prick tests can detect both antibody and cellular reactions. If your allergy is to certain grasses or trees, then pay attention to the cycle of pollination where you live. Staying in an air-conditioned building will help. In your home, be sure the air filters are kept clean. An electronic filter incorporated into a central heating and air conditioning system will help. If your allergies are to mold or fungi, beware that they can be especially heavy around grain, trees and other plants in the country. Summer cottages that have been closed for the winter can be full of mold.

If your allergy is to dust, you're probably allergic to dust mites. These multi-legged, microscopic creatures look like miniature wood ticks and ride on the dust particles floating in the air. You cannot see them. The real allergy is not to the dust or the mites, but to the mite droppings. Just as you cannot see the germs—the pathogens—that can be exchanged in a handshake or lurking in the water you drink, you cannot see mites riding magic carpets of dust floating in the air you breathe. Careful vacuuming, cleaning, and good air filters can help relieve the discomfort of a dust allergy.

If your allergy is to penicillin, this should be noted in your medical records. It would also be helpful if you wore a medical alert necklace or bracelet, because an allergic reaction to penicillin can be fatal. As mentioned, this is also true if you're allergic to bee stings. You should not only wear a medical alert medallion but you should carry an insect-sting kit if you're on a hike or at a picnic where you might be around bees. Remember, the allergic reaction to a bee sting can happen quickly, and an immediate antihistamine treatment is required.

Treatment for Allergies

In cases of mild to moderate immediate allergies, antihistamines are given to counteract an allergic reaction by chemically blocking histamine and histamine-like chemicals from being able to cause their mischief—an explosion of histamine and histamine-like chemicals—an allergic reaction. If you have persistent allergies, you may be able to get shots to protect you from the discomfort and dangers. In order to detect the antigens to which you are allergic, your doctor will begin by injecting small amounts of the suspected allergens into your skin, use skin patches of assorted suspected allergens, or offer more definitive tests. If your skin turns red after a skin test, this tells the doctor that the substance injected, or on the patch, is causing an allergic reaction. After identifying the allergen, or allergens, your doctor will begin giving you a series of small injections of that same offending allergen in order to slowly build up an immune response in your system. As this small amount of allergens enters your body, the helper T cells and the suppressor T cells in the command center send out the normal defense forces and, at the same time, command the B cell plasma cells to make the regular antibodies. This immune army regiment is composed of "Immunoglobulin-G (IgG)." Immunoglobulin G antibodies do not react with allergens in the same way as IgE. The IgG does not carry the allergenic invaders to the basophil or the mast cells. Because they can react faster than the IgE, and quickly destroy the incoming invaders, they leave no allergens for the IgE to bind. Slowly, your immunological memory becomes expanded, and you will no longer suffer from reactions to that allergen.

While this chapter has reviewed the major explanations for hypersensitivity, there are several more battles within the immune system that will result in allergic reactions. One of these occurs when certain regiments of the antibody army form what is known as an immune complex-mediated allergic reaction. In this reaction, your antibodies join with the invading antigens to form a complex molecule called an "immune complex," a hybrid unit that can cause types of hypersensitivity reactions such as inflammation of the walls of the blood vessels, inflammation of the kidneys, or of the joints. This damage is triggered by the immune complex becoming struck, either to cells lining the blood vessels or in the blood filtration systems of the different organs. Deposits of this immunological "junk" attract neutrophils and macrophages, who interact with it, but while doing so also damage the surrounding tissues and cells. Such damage causes inflammation and, in many cases, severe organ damage.

Chapter Twelve
Immune System Nutrition: Feeding the Immune Army

If you are a trivia aficionado, you might enjoy the debate as to whether it was Napoleon Bonaparte, Emperor of France; Frederick the Great, King of Prussia; or Claudius Galen, the Greek physician in his role as the chief physician of the Roman Army, who was the first to declare that "an army marches on its stomach." Regardless of the author, the quote states the truth that in order for an army to be strong and victorious, it must be well fed. Likewise, your immune system's army will perform better when it is well fed. That translates to you being well fed, and that, in turn, means a balanced diet. While good nutrition is important for your overall health, it is absolutely necessary for your immune system. The foods that help you build a healthy body are the same nutrients needed to produce a healthy immune system. The problem is, when you take in food to feed your hungry body, those hungry organs—such as your brain, heart, and lungs are fed first—so, your immune system must wait its turn at the back of the buffet line.

The nutrients you eat are used first to feed your essential organs (brains and nervous system). Next they feed your heart and its blood transportation system. Next in line are your most vital organs—lungs, liver, stomach, kidneys—and after that, your muscles and skin. After they've all been fed, then the white blood cells of the immune system are nourished.

The Importance of Sleep

"All you need is a good night's sleep." That's good advice, but nobody told you why.

Sleep is your body's repair shop. When you sleep, your body resets all its stress clocks. Your body systems slow down during sleep, except your immune system. While your other organs are in a moment of repose, your immune system gobbles up all the nutrients and does its best work. When you sleep, and your body slows, your immune system is at its most active and moves to the front of the line for the essential elements needed to feed your white blood cells. Often, sleep is the best prescription for fighting infection. It is not by accident that you can go to sleep feeling ill and wake up feeling better.

Fuel for the Human Engine

When you chew a bite of food, you mix it with saliva and grind it to a pulp so that you can swallow it. Your spittle contains certain enzymes which initiate the breakdown process. In your stomach, acid and digestive enzymes reduce the food into a liquid and some fibrous material. This is digestion.

The liquid then passes into your small intestines, where the nutrients are absorbed through the walls lining the bowels and are then passed through the small blood vessels, called "capillaries," into your blood. Once in the blood, these nutrients are carried to all the cells of your body. This "food" provides the energy to operate your systems, including your immune army. This energy is measured in calories. Just as gasoline is the fuel for your car because it produces an internal explosive combustion to run the pistons in your engine, the nutrients from your food produce calories, the heat energy to operate your body. The fibrous food and undigested, or unused materials, pass into your large intestines, where water and minerals are absorbed before being passed out of your body as waste.

One calorie is measured as the energy it takes to heat one gram of water to one degree. Your body needs a certain number of calories daily to operate fully. Calories are needed for your mental and physical activities, as well as for growing and repairing damaged or injured body parts. A growing teenager active in school and sports needs far more calories than his or her grandparents, especially if the grandparents spend a lot of time on their recliners watching television or on a park bench feeding squirrels.

When the food you eat provides more calories than your body uses, then the excess energy supply is stored as sugar and fat, and you gain weight. Of course, the opposite is also true. When you burn more calories than you take in, then you lose weight. Ideally, you should take in only the number of calories that you burn. This balance is not easy to maintain. But there is more to nutrition than calories. Hand in glove with calories are the nutrients: Protein, carbohydrates, fats, vitamins, minerals, and water. You must balance calories and nutrients when planning a healthy diet. Some foods are very high in calories but low in nutrients. And guess what? These are usually the foods you love to eat most, like candy and potato chips. An unbalanced diet high in calories and low in important nutrients can affect your immune protection. Other foods, such as lean meats, poultry, and fish, peas and beans, whole grains, fruits, and vegetables, are all relatively low in calories and rich in nutrients. These should provide the mainstay of a balanced diet. Packaged and processed foods are usually labeled with the number of calories they contain, along with a list of their nutrients. Many books and magazines provide a great deal of information about calories and nutrients for a wide range of foods.

Let's take a look at the different types of nutrients:

PROTEINS – THE KEYS OF LIFE

A protein is a molecule so complex that it is often referred to as a macromolecule. A protein molecule is made up of carbon, hydrogen, oxygen, and nitrogen. You probably know that two atoms of hydrogen and one atom of oxygen together form a simple molecule, water. On the other hand, carbon, hydrogen, oxygen, and nitrogen can come together as an organic molecule called an "amino acid." With this molecule, we have the molecule of life. Think of one amino acid molecule as a round bead—like the multicolored plastic beads that can be snapped, or popped, together to make a children's necklace—that can be linked with other amino acid pop beads. If we linked these beads together, we would have a "peptide." If we linked the peptides together in a long necklace-like chain, we would have a "polypeptide." Then, if we folded and twisted the polypeptide chain so that it corkscrewed and piled together, we would have a "protein" molecule.

The sequence in which the beads are put together and folded onto each other determines, as a start, the kind of protein molecule they will become. For example, if each bead were a different color, such as red, blue, green, or yellow, then the order that they were put together would be important. There are twenty different amino acids that can be in a chain, so we must picture twenty different colors. And then, they can occupy different positions, patterns, and sequences within the chain. While the number of amino acid pop beads in a protein varies, it is their shape that is important; they can be piled, twisted, and folded into an almost limitless number of shapes. Then, it is easy to understand how there can be billions and billions of different combinations of proteins.

You, of course, begin the process by chewing and swallowing. Minced and softened with saliva, food travels to your stomach to be digested, where the proteins are broken down to their simplest form. By this process, the amino acids are separated and passed along in red blood cells to individual cells, where each protein molecule is pulled apart into its original amino acid components. After these amino acid "pop beads" pass into a cell, they are reassembled as new protein molecules, according to the needs of that cell. All of this explanation is to point out that you must have the right kind of protein to stay healthy. Sadly, almost one-third of the people in the world do not get enough protein-supplying food. In the United States, too little protein is seldom a problem – excess protein is often converted and stored in the body as fat, and sometimes results in kidney and bone problems.

The extremely complex molecules are a major part of every cell. In your immune system, they form your antibodies and the structure of all white blood cells. When you lack protein in your diet, your immune system cannot function as it should.

There are twenty-two types of amino acids. Of these twenty-two types, thirteen can be manufactured within your body, while nine must be supplied by the food you eat. These nine are called "essential amino acids." Of the nine essential amino acids, phenylalanine, tryptophan, and valine are the amino acids essential for the production of your life-protecting antibodies.

What should you eat to get the necessary proteins? Meats and fish contain all of the essential amino acids. However, you can get a full set of essential amino acids from vegetables, nuts, legumes, grains, seeds, and dairy products, as long as you combine these so that all of the essential amino acids are eaten together. For example, a meal of rice and kidney beans, or peanut butter on whole wheat bread, would provide most of the essential amino acids. There are many such combinations that work.

Carbohydrates – The Major Fuel

While proteins are the building blocks, "carbs" provide one of your body's primary energy sources, the power to run your body's operating systems. If you do not get enough carbohydrates, then your body will borrow the proteins from other sources, such as your immune system. This results in a lowering of immune protection.

A carbohydrate molecule is made up of carbon, oxygen, and hydrogen. These molecules range in size and complexity, from small and simple to large and complex. The primary source of carbohydrates is from plants. As the molecules enter your digestive system, enzymes take a chemical and electrical sledgehammer to break the bonds holding the three elemental atoms together. As we have learned from atomic bombs, releasing bonded elements, here oxygen, hydrogen, and carbon, generates energy. It is this energy, the calories, that gives us life. The amount of energy released is determined by the number of atoms joined together in the carbohydrate molecules. The more complex the molecule, the longer the enzymes take to break it down. The energy from a complex molecule lasts longer than the energy from a simple molecule.

Sugars, starches, and cellulose are different types of carbohydrates. Sugars are the simplest carbohydrate molecules, and, therefore, are burned up quickly by your body. Sugars, such as those found in candy bars, are a quick energy source. Starches, such as spaghetti and potatoes, are complex carbohydrates that provide sustained energy. Cellulose is a fiber. The carbohydrate molecules of cellulose do not dissolve in digestion. Instead, they pass into and through your intestines like a great cleansing brush, removing undigested, and unused, food from your body. These carbohydrate molecules help keep your body clean and prevent a buildup of bacteria. Carbohydrates can also bind with proteins to form new molecules called "glycoproteins." This combination helps proteins, manufactured in cells, to pass through the cell membrane and be secreted. Antibodies are glycoproteins, and the sugars in the antibody tail function and not only to help the antibodies be released from the plasma cells but also aid in enabling the antibodies to bind to cell receptors.

If you make a point of eating whole grain breads, cereals, and legumes rather than refined starches such as white bread, cereals, or pastas made with white processed flour, you'll be getting cellulose along with the other two types of carbohydrates. Cellulose is also found in most fruits and vegetables.

Fats – Gotta Have 'Em and They Taste Good, Too

Fats are essential as cell membrane makers because they are needed to build the walls of all of your cells, including the white blood cells. Fats provide the needed cushion in your body. A layer of cells stuffed with fat provides the cushion between your skin and your muscles. Every one of your cells is set in fat, like bricks in mortar. Also, fats, like carbohydrates, provide energy – calories. When fats burn along with carbohydrates, they greatly increase your energy output. They also store energy so that when your body needs energy, it is there. This is why arctic explorers always carry large amounts of fatty foods to help keep their energy levels high.

Without fats, you would have to eat enough carbohydrates daily to provide the whole day's energy supply. Those times when you were short on needed carbohydrates, your body would reach in to take the needed carbon, oxygen, and hydrogen atoms from your protein supply, and as we noted before, this would weaken your immune system. Individuals on low-carbohydrate diets can protect their immune systems by eating green vegetables such as spinach and broccoli, as well as fatty fish like salmon, sardines, and tuna. Fats derived from plants like avocados, nuts, and olives, as well as uncooked olive oil, are also beneficial in maintaining a strong immune system while lowering your intake of carbohydrates.

Fats also have another job, and that is to absorb, and transport, the fat-soluble vitamins – A, D, E, and K. The reason most people like fat is that fat makes foods taste better.

Fat molecules contain the same three elements as carbohydrates—carbon, hydrogen, and oxygen—but in different proportions and they are arranged differently. The number of hydrogen atoms that combine with the carbon atoms determines whether the fat molecule is "saturated" or "unsaturated." The more hydrogen molecules, the more saturated the fat. The difference in the chemical nature of saturated and unsaturated fat molecules makes them behave differently.

You can easily see the difference: At room temperature, saturated fats, such as butter, are firmer and more solid than unsaturated fats, which are softer or liquid, such as vegetable oil. The difference affects how your body digests the fat. The unsaturated, liquid form of fat is more easily digested than the saturated fat. Much unsaturated fat is stored in the body, producing "cholesterol," a chemical essential to the health of cells. Many people think of cholesterol as bad, but in the correct proportions, it is essential to good health. Cholesterol is a building block of cell membranes, "myelin" (the fatty insulation for all of your nerve fibers), natural steroid hormones, and bile (the liquid that helps with digestion). In short, eating the right amount of fats, particularly unsaturated fats, along with carbohydrates and proteins, is important to maintain peak health and a vibrant immune system.

On the other hand, too much fat, particularly saturated fat, can lead to trouble with a buildup of fatty deposits, leading to the clogging of arteries and heart disease. Fats from animal foods, such as butter or the fat on meat, are mostly saturated fats, while fats from plant sources, such as corn or olive oil, are usually unsaturated.

Vitamins and Minerals

We cannot complete our understanding of nutrients without discussing vitamins and minerals. These components, known as micro-nutrients, are essential for maintaining a healthy, functional immune system and are acquired either through a healthy diet or by adding them as supplements. A vitamin is a chemical that your body uses as a catalyst. A catalyst is a chemical that helps other chemicals—proteins, carbohydrates, and fats—combine without changing, or being involved in, the resulting chemical combination. Basically, that means that they help the other nutrients—proteins, carbohydrates, and fats—do their job. Some vitamins break down the various elements of the nutrients so that those same elements can be reassembled in the different forms needed in the body by the different cells, including the white blood cells of the immune system. A lack of vitamins can lead to many different diseases.

There are thirteen life-essential vitamins. These are divided into two groups: Fat-soluble and water-soluble. When you eat more water-soluble vitamins than you need, they are simply washed out of your body as waste. On the other hand, the fat-soluble vitamins can be dissolved only in hydrogen-rich fats. As a result, these vitamins become stored in your body fat. If you take in more of these vitamins than you need, you could risk serious health problems. A balanced diet normally provides all of the vitamins necessary to keep your immune system in good working order. However, when taking vitamin supplements, care should be taken not to exceed the recommended higher levels for each vitamin.

Vitamin A is a fat-soluble vitamin that can be an efficient booster of the immune system. It can interfere with and prevent infections by enhancing the activity of T cells, B cells, and cytokines. Additionally, it helps maintain the integrity and function of skin cells and other cells lining the airways, digestive tract, and urinary tract, thus strengthening the barrier against infectious invaders, particularly bacteria and viruses. Concentrations of greater than one thousand micrograms in adults can lead to a disease called "hypervitaminosis A," which is vitamin A toxicity, and can cause bone pain, nausea, and skin problems.

Vitamin C is a powerful antioxidant that aids in immune function by enhancing the production and vital functions of white blood cells. Vitamin C can also lessen the severity of colds, flu, and other respiratory infections. It functions as a regenerative agent for Vitamin E, which is also another antioxidant. Care should be taken regarding overdoses of vitamin C.

Concentrations greater than two thousand mg can cause diarrhea, nausea, and gastrointestinal disturbances.

Vitamin D aids in calcium absorption, which is helpful in maintaining bone health and reducing the risk of autoimmune diseases. Some studies have suggested that a lack of vitamin D can lead to cancer, diabetes, and cardio-vascular disease. Your skin, when exposed to natural sunlight, can produce vitamin D. As with other vitamins beneficial to your immune system, excess vitamin D (greater than ten thousand International units/day) can cause problems such as a buildup of calcium in your blood, which can induce nausea and vomiting, and is usually associated with excess vitamin D coming from supplements. Prolonged excess can lead to bone pain and the formation of calcium, leading to kidney stones.

Vitamin E is a powerful antioxidant and it can help the immune system in fighting infections. It helps to boost immunity and reduce the risk of developing infectious diseases. Although overdoses of vitamin E are rare, they do occur when you take greater than one thousand mg/day on a regular basis. In such cases, studies have found that excess vitamin E can interfere with blood clotting and cause hemorrhages.

Vitamin B6 is essential for maintaining healthy production of red blood cells and plays a role in the production of antibodies. A lack of vitamin B6 can affect the healthy maturation of both T and B cells, as the vitamin is required for protein breakdown and healthy cell growth. Very little has been recorded about the effects of excess vitamin B6, usually caused by excessive use of supplements. Research has shown that excess levels lead to toxicity, causing loss of muscle control, painful skin lesions, heartburn, nausea, and sensitivity to sunlight. However, these symptoms only arise when doses greater than two hundred mg/day are taken. Minerals control the water and chemical balance in your body. Your immune system will respond to a mineral imbalance just as it would to a vitamin imbalance.

A mineral imbalance usually shortchanges the production of antibodies, reducing the immune army's rocket battalions. Minerals are divided into two types: Macro and micro minerals. The macro are the most important—calcium, sodium, chloride, potassium, magnesium, and sulfur—while only slight traces of the micros are needed—iron, zinc, selenium, iodine, copper, chromium, manganese, molybdenum, and fluoride. Of these, zinc is the most important for the immune system. Zinc is a trace mineral that is essential for immune function as it ensures that the white blood cells function properly. Zinc helps the white blood cells fight common infections, and it is also important in wound healing. Although essential, a high dose of zinc can also cause immunosuppression because the mineral can be toxic to the white cells in high concentrations. This does not mean that you can boost your immune protection by taking a zinc supplement. Many supplements, like Zicam, claim that the product (containing zinc) is beneficial for enhancing immunity, but such claims have not gained Food and Drug Administration (FDA) approval. This is true for most, if not all, zinc supplements. Remember, too much zinc can be harmful. Adverse effects of high zinc intake (greater than 150 mg/day) include nausea, vomiting, abdominal cramps, diarrhea, and severe headaches.

Selenium is perhaps the most important immunity-boosting mineral. It is essential for the optimal functioning of the innate immune system, especially macrophages and neutrophils. Additionally, selenium binds to certain proteins to form specialized proteins that play an important role in the creation of functional T cells. A deficiency of this mineral can impair, or inhibit, the production of active antibodies.

Magnesium is an important mineral because it co-functions with calcium to improve many cellular functions. It regulates the concentrations of vitamin D and zinc in the body. Magnesium also helps regulate inflammation, and a deficiency of this mineral can lead to chronic inflammation and allergies.

Iron is essential for red blood cells to transport oxygen to the cells of the body as it promotes the formation of hemoglobin. Iron is also involved in maintaining the correct functions of neutrophils and lymphocytes.

Copper augments the innate immune system by assisting in the destruction of bacteria and viruses. It is also important for the utilization of iron by the body. One of the most important functions of copper is to ensure that cellular enzymes work correctly, including cells of the immune system.

A balanced diet can provide zinc and all of the other vitamins you need – whole grains, lean meats, eggs, legumes, liver, and seafood are good sources of your needed minerals. In a word of caution, the taking of supplements for vitamins or minerals should be modulated with the advice of your health care providers, as unfortunate side effects may occur, especially if the dosages on the warning labels are ignored.

Please Pass the Salt

In medieval times, salt was often the measure of wealth, in addition to health. When the lord of a castle and his guests and attendants gathered at the great table, people of high rank sat "above the salt," while those of lesser station sat "below the salt." Social position was indicated by salt. It was traded as a commodity, and in some circumstances, it was considered as valuable as gold. Roman soldiers received part of their pay in salt, their *salarium* – the Latin origin of the word "salary."

Salt is the compound that results when atoms of sodium and chlorine combine. Both elements are chemically, highly active. Dissolved in water with trace amounts of potassium, salt separates into sodium and chlorine "ions," or electrically charged atoms. The sodium ions carry a positive electrical charge, while the chlorine ions carry negative charges. These particles in water are called "electrolytes," a solution through which electrical current can flow. Similarly, electrical currents flowing within your body through nerve impulses and ionic flows are important for many different body functions, including immune system transport of molecules and cells throughout the body. We generally add salt to our diet when cooking. In this case, there are two different types of salt available: Common table salt, which has iodine added, and sea salt, which is sun-crystallized sea water and does not contain iodine. Although sea salt often tastes better, table salt can be healthier.

Normally, if you eat a balanced diet and drink plenty of water, you will get enough salt and water. If you exercise vigorously, you then must drink more water to make up for the water loss from sweating. Muscle twinges and cramps are early warnings than your body is out of the proper salt-to-water balance.

This is followed not only by fatigue, but also by lowered immune protection.

You can survive for long periods of time without food, but only for a few days without water. A prolonged period without water leads to serious problems. Blood movement slows, lymph fluid thickens, and the immune system loses its main rivers for transportation. As a result, your body is more susceptible to an invasion by microbe invaders.

When your body loses water, the salt content goes up. A loss of water, called "dehydration," is not the same as perspiration, which not only removes both salt and water but also acts as a frontline defense against skin bacteria and other invaders. The loss of water increases the concentration of salt, which can lead to the death of cells, and when your cells die, so can you.

Chapter Thirteen
The Mind-Immunity Connection: You Gotta Laugh

Do you think that psychoneuroimmunology is a laughing matter? Well, it is, or at least part of it is, because, as explored in this chapter, laughter can help boost your body's immune response. This suggests an answer to the question: Does your mind play a role in the operation of your immune system? The answer is "yes." The connection between your mind and your immune system is called "psychoneuroimmunology," a fancy name given to a developing science promoted by Dr. Steven Locke, of Harvard University, and Dr. Bernie Siegel, of Yale University.

Your immune system functions through a series of commands from your brain. Electrical impulses from your brain flash along your nerves, and hormones carry chemical instructions throughout your body, and now we are learning how cytokines carry their protein messages from your brain across the blood-brain barrier to your immune system. Also, there are now studies indicating that, through cytokines, your immune system can talk back, sending its messages and even commands, to your brain. With this exchange of information and directions, your immune system tends to its job. If your brain commands it, your immune system activities can be increased--more white blood cells can be produced to attack an invading pathogen. It is as if your brain is an amplifier for the music of an immunological songstress. Laughter is often your immune army's marching band or maybe its theme song.

Professor Carmine Pariante, in a paper published in 2015, suggested the term "immunopsychiatry" as her label for the concept that the brain is not necessarily in charge but that some human behavior is governed by immune mechanisms from the central nervous system giving commands to your brain. This new science is just budding, but it could help explain some of the mysteries that exist in immune-brain interactions.

Over the centuries, healers and witch doctors have used spells and rituals to convince people who were sick that they, the healers and witch doctors, as wonder workers of their tribe, were possessed with magical powers and had special abilities to heal all ills and cure all diseases. Tribal medicine men achieved positions of great status in their tribes and communities because they were often successful in curing the sick or healing the injured.

Even though their knowledge of medicine was limited, these healers relied on an important key to health: the mind-body connection. What was, and still is, important, is that the weak and ill, and the injured, believed that the medicine man would make them well. Often, their faith in the healer is what counted. There is a powerful connection between your thoughts and beliefs and your health.

When a tribal healer treated a patient with a ritual, the patient believed he would be cured. His brain sent the upbeat message that did, in fact, increase the production of disease fighting white blood cells and this helped make him well. Although people have relied on "magic" healing for over five thousand years, it has only been in the last few decades that medical scientists have been able to demonstrate that you can turn your immune system on, or off, up or down, by how you think.

This idea was first demonstrated in the laboratory in 1975 by psychologist Dr. Robert Adler and an immunologist, Dr. Nicholas Cohen, at the University of Rochester in New York. The two doctors were testing a new drug to determine if it would be effective in suppressing immune systems. They set up their experiment by giving the proposed medicine to laboratory rats. To get the rodents to drink the drugs, they sweetened the experimental doses with rose water. The rats drank the sweet liquid eagerly. The drugs in the sweetened water worked and the rats' immune systems shut down. But later, when the doctors gave the same rats only rose water without the experimental drugs, their immune systems shut down just the same as if they were consuming the medicine.

This experiment resembled the earlier, and groundbreaking, work by Ivan Pavlov, the Russian physiologist who won the Nobel Prize in 1904 when he proved that the mind could affect the body's responses. Pavlov trained his dogs to understand that whenever he rang a bell, dog food would be served. Normally, at the smell and sight of food, the dog would begin to salivate. Over time, the dog learned to connect the ringing of the bell with the arrival of his food. Soon, his dog was salivating whenever the bell rang, without any food being put before him. Pavlov called this a "conditioned reflex," and it was the first proof of the mind-body connection.

The accidental experiment by Dr. Adler and Dr. Cohen, likewise, proved that there was a mind-connection to the immune system. We have since been able to confirm that the brain does communicate with the immune system, issuing electrical and chemical commands; and perhaps of greater importance, the new knowledge that cytokines, crossing the brain barrier, are extremely instrumental in your brain's ability to directly focus the immune system components to battle incoming pathogens.

Attitude

If you are a happy, upbeat person, convinced of your good health and your ability to overcome any disease that comes along, then you are helping your immune system. By facing challenges with a winner's spirit, you can affect your body's immune regulation. In Chapter Nine, we discussed the role of the cytokines in communicating with your immune system. Remember that as this chapter is written studies on the role of mind-immunity interactions are in their infancy. Much is yet to be learned, but the possibilities are exciting.

There are a number of conclusions that can be made. A smile and an optimistic attitude will signal your immune system to produce more white blood cells. More white cells provide you with greater immune protection. And if you can handle a smile, just once in a while, in order to please your immune system, why not try actual laughter? As I prepared this chapter, I was recognizing a growing body of scientific investigation suggesting that laughter, pure unbridled cachinnation, actually improves the activities of your immune army. So maybe we should say that not only is the army of our immune system healthier when your immunological troops are well fed, but also your soldiers will always appreciate a good joke and a belly laugh. Obviously, laughter is an outcrop of the feel-good condition of your mental processes; it counteracts the effect of stress and sadness, which will suppress the efficiency of your immune system.

A certain amount of stress is a natural, and expected, part of life. Do you recall the first time you had to go up on a stage in front of a crowd of people to act or sing? Remember that hard knot in your stomach, or feeling a little sick, or how your palms were sweaty? Those were signs of stress – the anxiety that prepares your body for action.

The human body is designed for moments of temporary stressful intensity. It is often referred to as the "fight or flight" response. Stress is nature's way of organizing your nerves, muscles, blood, and organs into a protective mode. As you face a stressful or dangerous situation, your blood pressure goes up and your heart rate increases. Your digestive tract shuts down, and you may feel a little queasy. You feel a boost of energy as your body's chemistry turns proteins into glucose (a type of sugar) for quick action.

It all begins in the brain. When you recognize a stressful situation (through your senses – sight, hearing, feeling, and so on), these perceptions are conveyed to the hypothalamus. The hypothalamus sends hormone messages to your pituitary gland, which lies at the base of your brain. The pituitary gland, in turn, sends hormone messages to other glands, which signal your organs and muscles to prepare for action. One of the glands that receives these instructions is your adrenal gland. The adrenal gland produces a number of chemicals, including adrenaline and steroids, as part of the fight or flight response. Adrenaline raises blood pressure, and the steroids shut down immune activity to direct all your energy to fighting or fleeing.

When that happens, your body-parts are "pumped up," except the immune system that goes the opposite way. It slows down, and this leaves your body with a lower resistance to disease and infection while directing more food energy for your muscles to use. For short periods, there's no harm done; your body adapts easily. But if you go through a severely stressful situation, it lasts a long time, or such situations happen frequently, your health can be jeopardized. This is known as distress.

Frustration, anger, fear, anxiety, or depression can lower your body's resistance to disease. A patient who reacts with shock and then depression when confronted with the sudden news that he has a cancerous tumor might unintentionally shut down his own immune system and speed the growth of the cancer. Another patient faced with the same news reacts by saying and believing "I can beat this disease" can help his immune system in its battle with cancer.

One medical study showed that college students undergoing the normal stresses of final examinations were more susceptible to colds, flu, and other infections during the examinations. These students suppressed their immune systems enough to allow infectious agents to cause disease.

Sudden or Extreme Cold

Stress on your immune system can come from outside sources. One is cold temperatures. As warm-blooded animals, humans are designed to regulate their body heat, maintaining a constant degree of warmth. If you dash outside on a cold day without a coat, your body is immediately stressed. Adrenaline is released and your body chemistry changes, causing your body temperature to rise. If you sit in a cold, drenching rain for a long time, the stress becomes distress, and those hormone messages fly into high gear. Your lymph nodes slow down, releasing fewer white blood cells. Your adrenal gland produces more hormones and increases blood pressure for more body heat. You are ready to battle the cold temperature, but there are fewer T cells, B cells, or antibodies to fight viruses or bacteria. If, unknown to you, your immune system had been quietly fighting off a bacterial, or viral, infection and you stayed very cold for a long time, your immune system might shut down long enough to allow the virus, or bacteria, to escape the soldiers of your immune system.

While getting cold will not, by itself, give you a cold, it can make you more susceptible. That's why when your mother said, "Bundle up, you don't want to catch a cold," she was on to something important.

Pain

Pain is the way the brain interprets messages from nerves that an injury, disorder, or disease is taking place. When the brain receives these messages, it issues commands. These commands instruct your nerves to direct an instant reaction. For example, you quickly jerk your hand back from a hot stove, or blink your eye when a gnat invades. You react to stop the pain. Your nerves may also direct some involuntary changes, such as an increased heart rate and faster breathing.

Pain also creates stress and puts you into the fight or flight mode. If pain is prolonged or constant, as with such diseases as arthritis, cancer, and liver and kidney disorders, then the stress of pain becomes distress, and the slowed down immune system increases the risk of other diseases. Depression and loss of sleep, triggered by chronic pain, also increase the likelihood of disease.

Sleep Is a Very Good Answer

As we mentioned earlier, "All you need is a good night's sleep." But we didn't explain why.

Sleep is the body's repair shop. When you sleep, your body resets its stress clocks. Your body systems slow down during sleep – except your immune system. Remember we told you that when the proteins are passed out, your immune system is at the back of the line? While all the other body systems slow down—major organs are at rest, heartbeat is slower, and breathing more relaxed—your immune system is first in line for proteins, vitamins, and minerals. Your amino acids are made available for the production of antibodies. During the first few hours of deep sleep, your pituitary gland sends out hormone messengers in the combined forces of the cytokine brigades we discussed in Chapter Nine to deliver the protein supply where it's needed to produce white blood cells, which in turn produce antibodies and other proteins associated with the immune system.

Often, sleep is the best prescription for fighting infection. It is not by accident that you can wake up cured of your illness. On the other hand, a lack of sleep, or even disturbed sleep, can slow recovery, or make you more susceptible to a disease.

Sleep occurs in stages. One important stage is called "REM sleep," which stands for rapid eye movement; another is called "NREM, or deep sleep," which stands for non-rapid eye movement. REM sleep is characterized by dreams and a flickering of the eyes back-and-forth. It is the deeper sleep, NREM sleep, during which your blood pressure drops, your heart rate slows down, your breathing becomes slower, and your body temperature falls. During the night, you pass in and out of these stages several times.

Each stage of sleep has its own functions, and experiments have demonstrated that not getting enough of one, or the other, type can have serious consequences. In REM sleep, your body is resting but your brain remains fully active. This is the period of active dreaming. Dreaming is not completely understood, but some scientists believe that dreaming is the way your brain sorts out sights, sounds, emotions, and thoughts from the day and organizes what should be kept in your memory, where it will be stored, and what should be disregarded, or forgotten.

In NREM sleep, all energy requirements are lowered. The immune system functions at its highest level, sweeping the body of invaders. In experiments where volunteers were allowed only REM sleep and not NREM sleep, the volunteers became sluggish and depressed. If people do not get NREM sleep for a prolonged period, they can suffer severe psychological damage. You need both REM and NREM sleep to stay healthy.

As discussed in an earlier chapter (Chapter Five), when the immune system remembers that it has met a certain invader before, it searches through its inventory in the lymph node for an identified antibody producing B cell that it can call to send the correct antibodies to attack the invader. If, on the other hand, a bacterium or virus invades your body for the first time, it can take many days before the immune system can process this information and direct the B cells to develop the necessary antibodies to destroy the invader. As a result, you may be sick for a number of days before you begin to recover. During part of the time you are sick, you might have a fever. When your immune system sends urgent messages to the brain to "please help until the needed antibodies can arrive in sufficient force to destroy the attacking viruses," your "hypothalamus"—the part of your brain that controls body heat, among other things—increases your body temperature. This increase in body temperature—the fever—helps to kill the attacking pathogens and increases the ability of macrophages and killer T cells to come to the site of the invasion. It is the cytokine messengers (IL-1 and TNFα) that successfully perform their Paul Revere type of urgent, and essential, messenger service.

Can the Brain Take Charge?

The separation between the brain and the rest of your body is walled by a barrier known as the blood-brain barrier. Prior to the 1950s, it was believed that the white blood cells of the immune system did not, and could not, cross this wall. That is no longer the case. The remarkable discoveries, as discussed in the prior chapter, reveal that those magical proteins, the cytokines, do find service across the border. In the 1960s, scientists discovered that the brain was sending "hormones" and other chemical messengers, called "neurotransmitters," to exercise certain control functions of the immune system. It was then believed that control was part of the automatic functions of the brain, directing certain immune functions through the system of direct signals, through the nervous system, and chemically, through the hormones and neurotransmitters. But, before the twentieth century was finished, the men and women of science had to take another look: along came the fascinating cytokines. The more that the scientists learned, the more, life-saving discoveries are in our future.

In the early 1980s, experiments revealed that activated white blood cells in the skin, stomach, or throat, for example, could send cytokine messages to the brain through the blood or through the peripheral nerves with cytokine receptors. There was more to come.

Now, well into the twenty-first century, the science of brain-immune system connection is still in its infancy. We now know that certain white blood cells, which are not found in the brain, and cannot cross the blood-brain barrier, can, however, when activated, actually cross that forbidden wall, and activate certain cells in the brain—called "microglia"—to search for invaders, and then destroy the enemy. The messages, the exchange of information, the exchange of commands, through the magic of cytokines makes the brain-immune system a two-way communication network. This information, then, raises a question: Does the immune system exercise any control over the brain? The answer appears to be "yes." Cytokines from white blood cells induce the hypothalamus to raise body temperature during fevers and inflammation, and cytokines from microglia can activate white blood cells. The understanding that there exists bi-directional communication has given rise to the introduction of a new field of study as proposed in 2014 by G.M. Khandaker and colleagues in an article in "Lancet Psychiatry," in which they introduced the term "Immunopsychiatry." The concept of a bi-directional set of communication mechanisms raises the hope that some unwanted behaviors might be able to be affected by controlled immune system regulation. Interestingly, a 1982 study by Harvard professor Herbert Benson, working with monks in Tibet, found that meditating monks could exert mental control over their own bodies so that they could raise the temperature of their fingers and toes.

All of this information leads to the next question: can you consciously and deliberately control your own immune system to fight disease? While this area of study has yet to provide real, scientifically proven, answers, a look into the future suggests that the answer will, one day, be "yes."

Chapter Fourteen
Immunotherapy:
A Look at the Future

With new insights and with new ways to enhance our already marvelous immune system, we welcome "immunotherapy." Under this title, many new approaches to enhancing the immune system have been developed and are being researched.

Monoclonal Antibodies

First consider "monoclonal antibodies", as you will recall from earlier chapters, antibodies are the Y-shaped proteins that are produced by the B cells (white blood cells) to target specific invaders as identified by other white blood cells. This form of immunotherapy has been in development since the early 1970s when the technique of producing monoclonal antibodies was introduced by Drs. Georges Köhler and César Milstein, who received the 1984 Nobel Prize for their work. So far, this immune system enhancement has been used in the treatment of a range of disorders such as arthritis, allergic asthma, psoriasis, and Monoclonal Antibody IV Therapy has been highly effective against COVID-

19. Even before COVID-19 monoclonal antibodies provided significant treatment of some deadly cancers. Cancers being your own cells gone rogue multiply and replace their brother and sister cells. When this happens, the cancer cells act as new, hostile invaders from within. When a cancer becomes antigenic, it will trigger cytotoxic T cells and antibodies to attack. When they attack, both the specific T cells and the antibodies locate the antigenic molecular markers on the cancer, attach to the markers, and begin the destruction of the cancer. Sometimes this is enough, and cancer never gets to fully develop, and fortunately, you will have never known you had been under attack. However, when the cancer grows faster than the immune system can respond, the answer may come with engineered monoclonal antibodies, the new, synthetic, but potentially effective warriors.

These new antibodies are made by removing the cancer cell's antigenic markers and injecting them into white mice. Through a complex process, antibodies to the cancer markers are produced. From these, antibody producing cells, called "hybridomas," become the final product. When a large colony of hybridomas has been collected, these are then injected into the patient, where they will produce literally tens of billions of antibodies that, in turn, will attack <u>only</u> the identified cancer. When these antibodies attack, they will not only begin the destruction of the cancer, but with the aid of the cytokines, will call-in other members of the white blood cell group to the cancer for its destruction. The scope and use of monoclonal antibodies was greatly increased in 1988 by the work of Dr. Greg Winter and his team, who developed a way to "humanize" monoclonal antibodies, thus removing potential immune reactions against the mouse part of the antibodies. Monoclonal antibodies, when produced in a similar manner to attack flu viruses, will, by the volume of their overwhelming immune system army, overpower even a fast-expanding viral attack and win the battle.

Monoclonal antibodies have demonstrated success in treating other diseases and disorders as the medical experts look to future progress in the value of monoclonal antibodies.

Adoptive Cell Therapy

In an attack on a wide range of tumors found in internal organs such as the stomach, esophagus, and ovaries, a search is made of the mutations found within the tumor. The solid, hard tumor is removed surgically and examined to determine the mutations in the tumor that were responsible for its development. Then the white blood cells—the "tumor infiltrating lymphocytes" (TILS)—are removed and collected. These very specific white blood cells that had been attacking the tumor but were not winning, are then multiplied by the tens of billions in the laboratory. After the patient is treated with chemotherapy, the patient's own multiplied, specific TIL cells are infused into the patient's bloodstream so that the tumor can be attacked by this complete army of "overwhelming force," tailored to attacking, and destroying, the identified tumor. This technique, still in the early stages of its development, has been described by Dr. Steven Rosenberg, of the National Cancer Institute, who has pioneered this procedure, as "The very mutations causing the cancer will be the Achilles' heel."

While this approach has not been effective in many of the patients in the trials, it is only the beginning. In the summer of 2018, a Florida woman whose breast cancer had spread throughout her body underwent this experimental treatment and was a lucky survivor, with complete elimination of her cancer. While not all recipients of the enhanced lymphocytes receive good results, the trials and development continue. This is an approach that holds promise for the future.

Nanobiotics

Chemotherapy drugs are used, traditionally, to kill fast-growing cancer cells. The problem with the use of chemotherapy has always been the collateral damage done to the patient's body – the unwanted side effects. Dr. Sylvain Martel at the Polytechnique Montreal has identified a bacterium, *magnetotactic cocci,* containing iron molecules that allow it to be directed on a compass course by the use of a magnetic field. A strain of this bacterium was found in a low-oxygen region of water in the Pettaquamscutt Estuary in Rhode Island. Since tumors tend to be heavy consumers of oxygen, they tend to be a natural attraction for the *magnetotactic cocci* bacteria. By coating the bacteria with a chemotherapeutic drug, injecting the bacteria into an area near the tumor, and assisting its travel by using weak magnetic fields and allowing the bacteria to then seek the low oxygen area, powerful drugs are carried to the cancer. Once again, this is working now in laboratories. Meanwhile, the bacteria cannot survive in the heat of a warm-blooded mammal and dies within thirty minutes; that is, it delivers the drugs and departs. Again, this is the beginning.

Other treatments include taking white blood cells (T-cells) from a patient's body, modifying the genes, and injecting these modified cells back into the patient for destruction of the patient's cancer. This has been recognized as one of the most important developments in cancer research. While genetically modified food products have raised wide controversy, similar techniques with the immune system have been hailed as major breakthroughs in cancer research.

Engineered Viruses – Phages

A "phage" is the shortened version of the word "bacteriophage." The idea of fighting bacteria with a virus was first successfully explored over one hundred years ago in Russia, where doctors developed a combination of several viruses, which they called bacteriophages, or phages. When seen in an electron microscope, the viruses resemble spindly-legged spiders with long necks capped with a twenty-sided polygon. Fortunately, these viruses are extremely fussy: they only attack specific bacteria. As new bacteria have been developing and mutating to become resistant to antibiotic drugs, new ways have been sought to combat the antibiotic-resistant bacteria. Because phages only attack certain bacteria, they, when engineered to target the newly mutating bacteria, perform as an important new tool. Scientists continue to work in order to engineer the phages to become new weapons against cancer.

By the Numbers

At the University of California, San Diego, in affiliation with Moores Cancer Center, a highly sophisticated computer program, I-Predict, analyzes the DNA of a cancer patient's cells and tumor characteristics, then searches all known and available treatments, anticancer drugs, and millions of drug combinations, chemotherapy availabilities, hormonal therapies, and even non-cancer drugs, to develop a proposed treatment, all looking for the best therapy for each individual patient. This is, again, a new approach. Guided by the concepts of artificial intelligence, there is a foreseeable future for precision medical-computer-program tools for effective, and customized, treatments against cancer, a disease where one size does not fit all.

And Then There Is Aging

As you take a look at the future of immune system enhancements, you must also consider what happens to your immune system as you age. We start with the chicken-and-egg issue applied to your immune system: which comes first? Will your immune system weaken as you age, or does a decline in the curative activity of your immune system contribute to your aging?

In 1964, Dr. Allan Goldstein, when he was a postdoctoral student, with his professor, Dr. Abraham White, in their laboratory at the Albert Einstein College of Medicine of Yeshiva University in New York, made a discovery that lifted the often-ignored human organ, the thymus, from obscurity to fame. The thymus is a gland beneath your breastbone which had been considered inconsequential until the doctors discovered in their experiments with mice that, indeed, the thymus was a prime organ in the processing of all of our immune functions; it was the processing organ for our T cells; the organ that delegated assignments for all the T-cells passing from our bone marrow. The discovery came from their isolation of small proteins that make up the hormones from the thymus, the "thymosins." It is the thymosins that we need. As we age, our thymus shrinks and transfers the T-cell control to our lymph nodes. Nevertheless, we need thymosins for T cell development and production. While the number of white blood cells in our bodies does not decrease, the wear-and-tear on our T cells begins in our sixtieth year. Thus, it appears that the health of our white blood cells is associated with robust aging and that the loss of T-cell regulation results in antibodies making mistakes—like an army of feeble old men bumbling around the battlefield—sometimes attacking our own bodies—autoimmunity.

Also, the loss of T-cell regulation allows cancer growth to take place. When immunological debris from immunological battles is not gobbled up and removed by macrophages, our aging bodies, unfortunately, allow the debris to organize into new body-hostile units called "immune complexes." These complexes deposit themselves in the kidneys, blood vessels, and other organs, creating a new set of diseases, including blood clots and resulting strokes, as well as kidney disease. The commercially developed artificial thymosins, hormones developed since 1964, along with other artificially developed hormones, have been successful in treating a host of disorders and offer hope for prolonging healthy lives. Who knows, perhaps the long-forgotten thymus maybe the real well-spring of the fountain of youth.

A Final Note

I like to believe that your journey through these pages has rewarded you with a new understanding of the magic within your body through newly acquired knowledge. But before I close the book on our visit together, I report that in the 1970s, Norman Cousins published in the New England Journal of Medicine his personal story of his non-medical, self-treatment, of his autoimmune disease, gaining complete remission with a reduction of stress through regular, mirthful laughter. This was one of the original suggestions that laughter could positively affect the immune system. The follow-up studies a decade later, by Dr. Lee S. Berk and D. Stanley Tan at the Loma Linda University Schools of Allied Health and Medicine, were able to confirm that hearty laughter enhanced the hormones in the endocrine system and decreased the levels of cortisol and epinephrine, which together increased the production of white blood cells. This was reported in 2010 by the Federation of American Societies for Experimental Biology.

Four years later, in a 2014 article in *Immunology*, the publication of the British Society of Immunology, there was a report of a little experiment that found that *some* patients enjoying relaxation, and particularly hearty laughter, had actually caused a meaningful increase in the production of their NK Killer T cells, regularly and predictably, spiking in intensity exactly twenty-four hours after the entertainment event that evoked the patient's raucous laughter. When this limited experiment was repeated, the result was the same. This was not a scientific study. This mind-immune system connection is still in its infancy. Some researchers suggest that there is no scientific evidence of a laughter meter for immune reactions, but we suggest that the concept is still an unproven theory and should not be rejected out of hand. Are the above events the visible tip of a new immunological iceberg of discovery?

There are more mysteries yet to unfold in the future for your and our marvelous immune system! That having been said, and this book having been read, you will hopefully treat your immune system with kindness, and await the promise of improved immunity through the mind-body connection. In the end, perhaps, after all, laughter could be the best medicine.

REMEMBER THAT, AND REPEAT AFTER ME:

One hundred million lymphocytes
Working days, but mostly nights
Engaged in antigenic fights.

Now start your psycho-imaging
Begin your T cells scrimmaging.
Close your eyes and catch the sights
Of tumbling, rounded shapely whites,
The romping, rolling leukocytes.
So gather 'round and listen closely
Imagine things now flowing ghostly.
In your mind, always the whites,
Your hundred million lymphocytes
Must win their immunological fights.
Then picture cancer lurking smugly,
Stretching grossly, looking ugly.
Catch him in his early stages,
Smite him with your macrophages.
Charged with cytotoxic rages.
Concentrate upon the thrills
Engendered by the neutrophils
Curing you of all your ills
Replacing all those nasty pills.

THE END

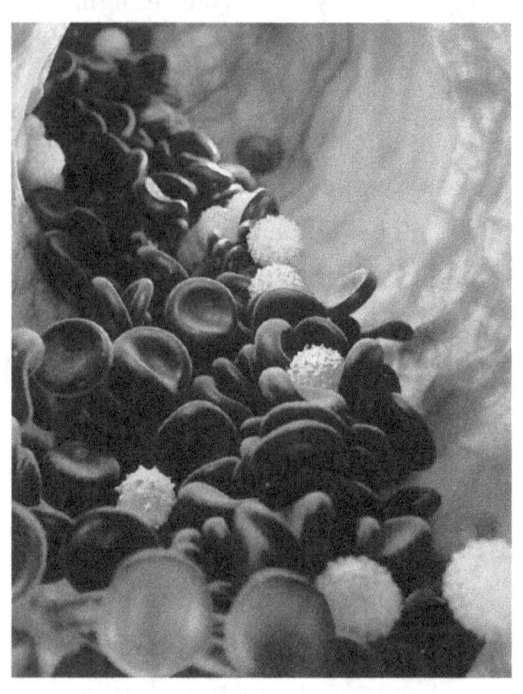

Red blood cells and platelets along with the white blood cells of the immune system flow through a blood vessel.

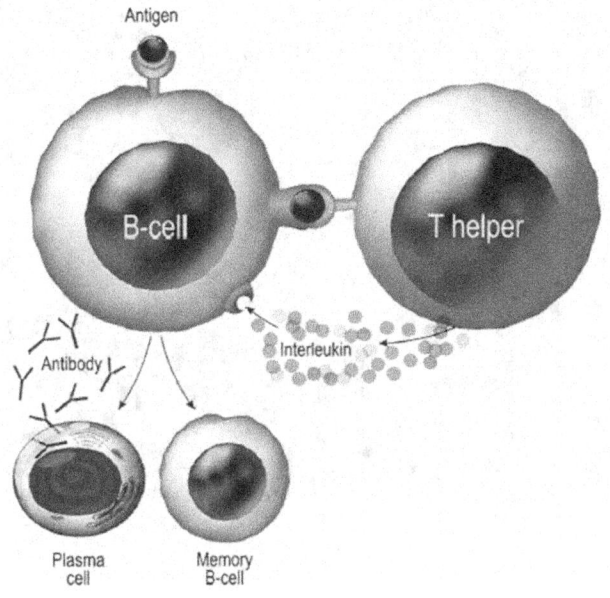

A diagram of the flow from T-helper cells to B-cells forming plasma cells for production of and storage of antibodies as aided by interleukin proteins (cytokines) messengers and regulators.

Plasma cell releasing manufactured antibodies.

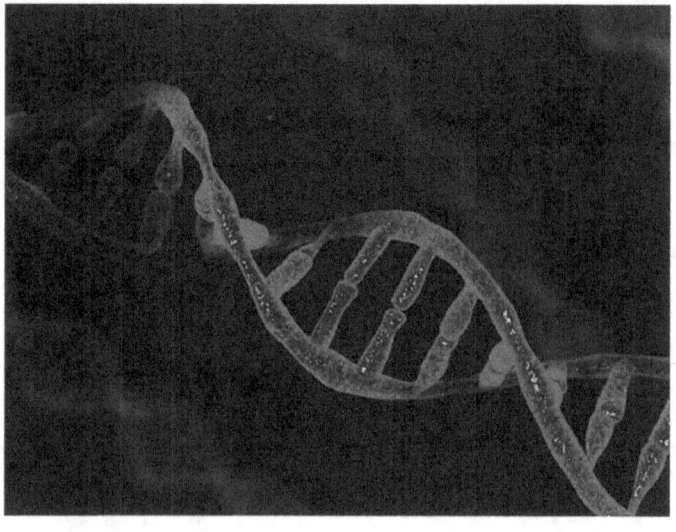

DNA. the double stranded helix of organic molecules found in all life forms.

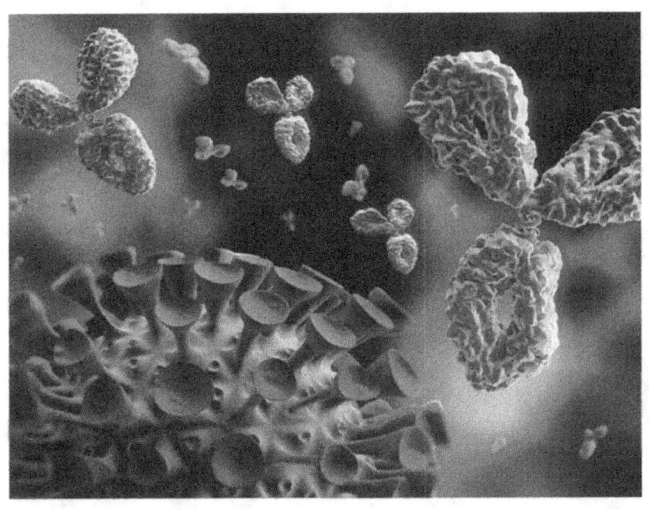

A illustration of antibodies mounting an attack on a COVID corona virus.

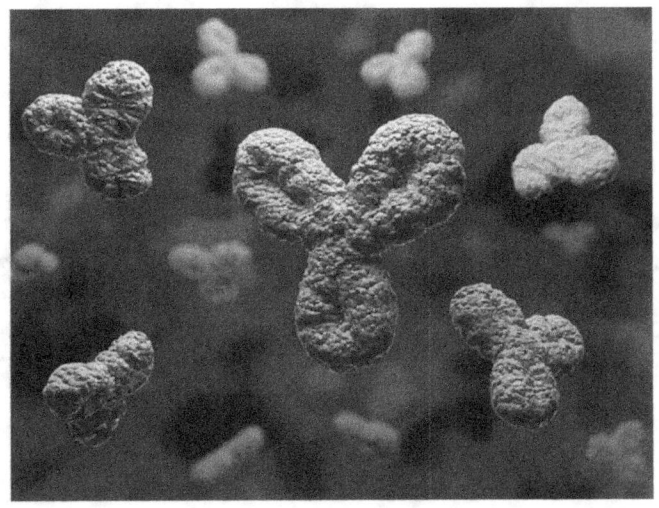

A illustration of antibodies showing the polypeptide assemblage of amino acids.

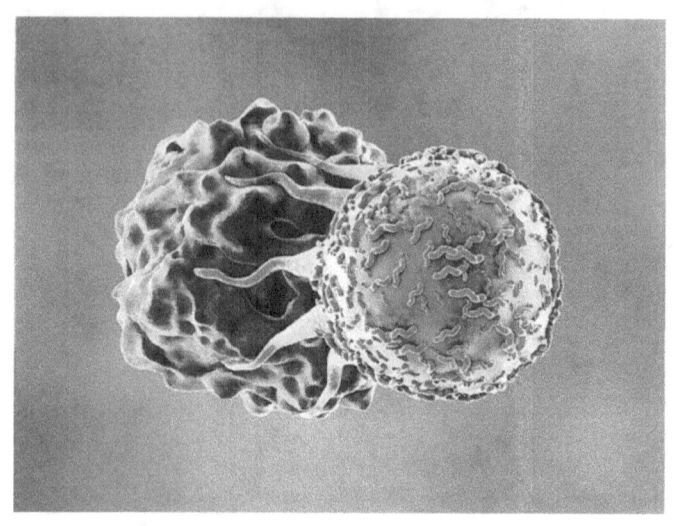

Killer T-cells attacking a cancer.

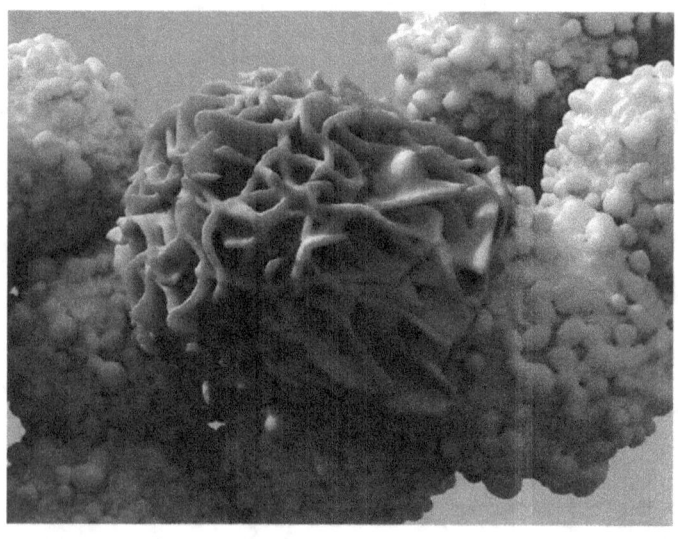

Large white blood cell (macrophage) devouring and destroying cancer cell.

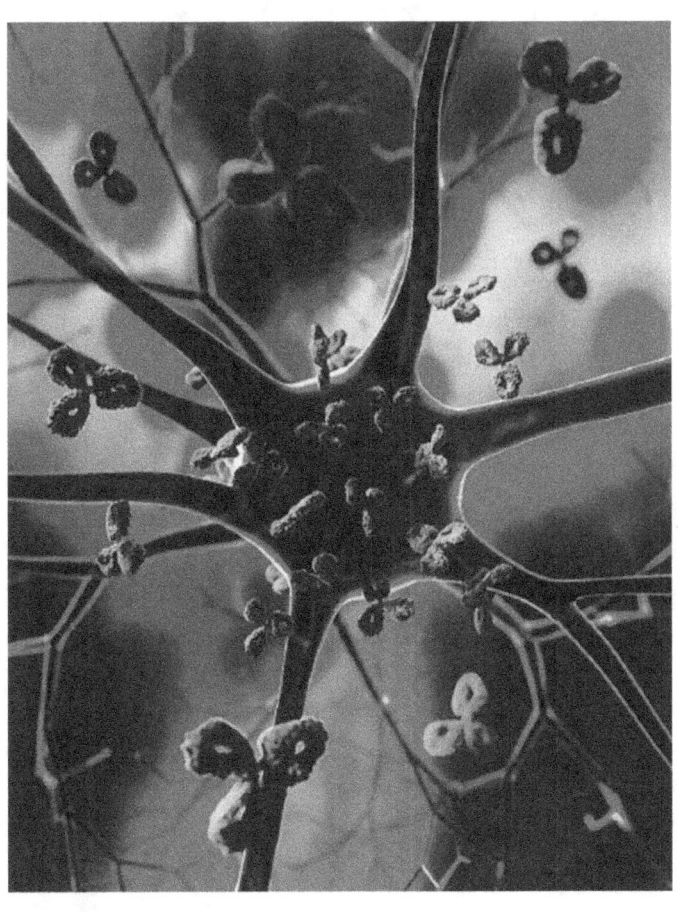

Antibodies attacking a neuron during an autoimmune attack such as in multiple sclerosis (MS)

www.ingramcontent.com/pod-product-compliance
Lightning Source LLC
Chambersburg PA
CBHW070636220526
45466CB00001B/197

9781685626204